T0214161

Lecture Notes in Computer Science   11826

More information about this series at http://www.springer.com/series/7412

George Bebis · Takis Benos · Ken Chen ·
Katharina Jahn · Ernesto Lima (Eds.)

Mathematical and Computational Oncology

First International Symposium, ISMCO 2019
Lake Tahoe, NV, USA, October 14–16, 2019
Proceedings

 Springer

Editors
George Bebis
University of Nevada
Reno, NV, USA

Takis Benos
University of Pittsburgh
Pittsburgh, PA, USA

Ken Chen
The University of Texas
MD Anderson Cancer Center
Houston, TX, USA

Katharina Jahn
ETH Zurich
Basel, Switzerland

Ernesto Lima
The University of Texas
Austin, TX, USA

ISSN 0302-9743 ISSN 1611-3349 (electronic)
Lecture Notes in Computer Science
ISBN 978-3-030-35209-7 ISBN 978-3-030-35210-3 (eBook)
https://doi.org/10.1007/978-3-030-35210-3

LNCS Sublibrary: SL6 – Image Processing, Computer Vision, Pattern Recognition, and Graphics

This Springer imprint is published by the registered company Springer Nature Switzerland AG
The registered company address is: Gewerbestrasse 11, 6330 Cham, Switzerland

Preface

It is with great pleasure that we welcome you to the proceedings of the First International Symposium on Mathematical and Computational Oncology (ISMCO 2019), which was held in Lake Tahoe, Nevada, USA (October 14–16, 2019).

The purpose of ISMCO is to provide a common interdisciplinary forum for mathematicians, scientists, engineers, and clinical oncologists throughout the world to present and discuss their latest research findings, ideas, developments, and applications in mathematical and computational oncology. Despite significant advances in the understanding of the principal mechanisms leading to various cancer types, less progress has been made toward developing patient-specific treatments. Advanced mathematical and computational models could play a significant role in examining the most effective patient-specific therapies. ISMCO aspires to enable the forging of stronger relationships among mathematics and physical sciences, computer science, data science, engineering, and oncology with the goal of developing new insights into the pathogenesis and treatment of malignancies.

The program included eight oral sessions, one special track, one tutorial, three invited talks, and seven keynote presentations. We received 30 submissions from which we accepted 19 submissions (7 papers and 12 abstracts) for oral presentation. This LNCS volume includes only the papers accepted for presentation; all abstracts accepted for presentation appeared in an online volume of Frontiers (link is provided on the ISMCO website).

All submissions were reviewed with an emphasis on the potential to contribute to the state of the art in the field. Selection criteria included accuracy and originality of ideas, clarity and significance of results, and presentation quality. The review process was quite rigorous, involving three independent blind reviews followed by several days of discussion. During the discussion period we tried to correct anomalies and errors that might have existed in the initial reviews. Despite our efforts, we recognize that some papers worthy of inclusion may have not been included in the program. We offer our sincere apologies to authors whose contributions might have been overlooked.

Organizing ISMCO for the first time was rewarding but also challenging due to the diverse background and interests of the targeted audience. Although significant advances have been made in various fields individually, it is evident now more than ever that new challenges in oncology can only be addressed by truly transcending disciplinary boundaries. Effectively bridging the gap among physical sciences, computer science, engineering, data science, and oncology is an absolute necessity in the hope of making significantly more progress in the fight against cancer.

Many contributed to the success of ISMCO 2019. First and foremost, we are grateful to the Steering, Organizing, and Program Committees; they strongly welcomed, supported, and promoted the organization of this new meeting. Second, we are deeply indebted to the keynote speakers who warmly accepted our invitation to talk at ISMCO 2019; their reputation in mathematical and computational oncology added significant value to and excitement to the meeting. Next, we wish to thank the invited speakers, the authors who submitted their work to ISMCO 2019 and the reviewers who helped us to evaluate the quality of the submissions. It was because of their contributions that we succeeded in putting together a technical program of high quality. Finally, we would like to express our appreciation Springer-Verlag, Frontiers, and the International Society for Computational Biology (ISCB) for sponsoring ISMCO 2019.

We sincerely hope that ISMCO 2019 offered participants opportunities for professional growth. We look forward to many more successful meetings in mathematical and computational oncology.

September 2019

George Bebis
Takis Benos
Ken Chen
Katharina Jahn
Ernesto Lima

Organization

Steering Committee

Anastasiadis Panagiotis	Mayo Clinic, USA
Bebis George (Chair)	University of Nevada, Reno, USA
Levy Doron	University of Maryland, College Park, USA
Rockne Russell	City of Hope, USA
Schlesner Matthias	German Cancer Research Center, Germany
Tafti Ahmad	Mayo Clinic, USA
Vasmatzis George	Mayo Clinic, USA

Program Chairs

Benos Takis	University of Pittsburgh, USA
Chen Ken	MD Anderson Cancer Center, USA
Jahn Katharina	ETH Zurich, Switzerland
Ernesto Lima	University of Texas, USA

Publicity Chairs

Loss Leandro	QuantaVerse, ITU, ESSCA
Scalzo Fabien	University of California at Los Angeles, USA
Erol Ali	Eksperta Software, Turkey

Local Arrangements Chairs

Nguyen Tin	University of Nevada, Reno, USA
Petereit Juli	Nevada Center for Bioinformatics, USA

Special Tracks Chairs

Dadhania Seema	Imperial College London
Marias Kostas	Foundation for Research and Technology – Hellas

Tutorials Chairs

Bhattacharya Debswapna	Auburn University
Shehu Armada	George Mason University

Web Master

Isayas Berhe Adhanom	University of Nevada, Reno

Program Committee

Pankaj Agarwal	BioInfi, USA
Bissan Al-Lazikani	The Institute of Cancer Research, UK
Max Alekseyev	George Washington University, USA
Panos Anastasiadis	Mayo Clinic, USA
Noemi Andor	Moffitt Cancer Research Center, USA
Ioannis Androulakis	Rutgers University, USA
Dinler Antunes	Universidade Federal do Rio Grande do Sul, Brazil
Ferhat Ay	La Jolla Institute, USA
Erman Ayday	Case Western Reserve University, USA, and Bilkent University, Turkey
Matteo Barberis	University of Surrey, UK
Anastasia Baryshnikova	Calico Life Sciences, USA
Ali Bashashati	University of British Columbia, BC Cancer Agency, Canada
George Bebis	University of Nevada, Reno, USA
Takis Benos	University of Pittsburgh, USA
Sebastien Benzekry	Inria, France
Debswapna Bhattacharya	Auburn University, USA
Jiang Bian	University of Florida, USA
Valentina Boeva	Institut Cochin, INSERM, CNRS, France
Mary Boland	University of Pennsylvania, USA
Matthew Breitenstein	University of Pennsylvania, USA
Amy Brock	The University of Texas at Austin, USA
Anton Buzdin	Omicsway Corp., USA
Raffaele Calogero	University of Torino, Italy
Emidio Capriotti	University of Bologna, Italy
Hannah Carter	University of California San Diego, USA
Aristotelis Chatziioannou	National Hellenic Research Foundation, Greece
Jake Chen	University of Birmingham Alabama, USA
Ken Chen	MD Anderson Cancer Center, USA
You Chen	Vanderbilt University, USA
Juan Carlos Chimal	Centro de Investigación en Computación del IPN, Mexico
Heyrim Cho	University of California, Riverside, USA
Giovanni Ciriello	University of Lausanne, Switzerland
Colin Collins	Vancouver Prostate Center, Canada
Francois Cornelis	Sorbonne University, France
Paul-Henry Cournede	Laboratory MICS, CentraleSupélec, France
Simona Cristea	Harvard University, USA
Seema Dadhania	Imperial College London, UK
Subhajyoti De	Rutgers University, USA
Francesca Demichelis	University of Trento, Italy
Mohammed El-Kebir	University of Illinois at Urbana-Champaign, USA
Peter Elkin	Ontolimatics, USA

Richard Levenson   University of California, Davis, USA
Doron Levy   University of Maryland, USA
Xiaotong Li   Yale University, USA
Ernesto Lima   The University of Texas, USA
Michal Linial   Hebrew University of Jerusalem, Israel
Doron Lipson   Foundation Medicine, USA
Hongfang Liu   Mayo Clinic, USA
Leandro Loss   QuantaVerse, ITU, ESSCA, France
Shaoke Lou   Yale University, USA
Brad Malin   Vanderbilt University, USA
Anna Marciniak-Czochra   University of Heidelberg, Germany
Kostas Marias   Foundation for Research and Technology - Hellas,
                                Greece
Scott Markel   Dassault Systèmes BIOVIA, USA
Florian Markowetz   University of Cambridge, UK
Schaefer Martin   EMBL/CRG Systems Biology Research Unit,
                                Centre for Genomic Regulation, Germany
Bernard Moret   Ecole Polytechnique Fédérale de Lausanne,
                                Switzerland
Abu Mosa   University of Missouri, USA
Henning Müller   HES-SO, Switzerland
Ulrike Münzner   Kyoto University, Japan
Radhakrishnan Nagarajan   University of Kentucky, USA
John Nagy   Arizona State University, USA
Tin Nguyen   University of Nevada, Reno, USA
Layla Oesper   Carleton College, USA
Baldo Oliva   University Pompeu Fabra, Spain
Laxmi Parida   IBM, USA
Bahram Parvin   University of Nevada, Reno, USA
Raphael Pelossof   MSKCC, USA
Shirley Pepke   Lyrid LLC, USA
Juli Petereit   University of Nevada, Reno, USA
Evangelia Petsalaki   EMBL-EBI, UK
Martin Pirkl   ETH Zurich, Switzerland
Angela Pisco   Chan Zuckerberg Biohub, USA
Clair Poignard   Inria Bordeaux-Sud Ouest-Team MC2, France
David Posada   University of Vigo, Spain
Gibin Powathil   Swansea University, UK
Natasa Przulj   University College London, UK
Teresa Przytycka   NIH, USA
Víctor M. Pérez-García   Universidad de Castilla-La Mancha, Spain
Ami Radunskaya   Pomona College, USA
Luke Rasmussen   Northwestern University, USA
Johannes Reiter   Stanford University, USA
Katarzyna Rejniak   H. Lee Moffitt Cancer Center & Research Institute,
                                USA

Isidore Rigoutsos	IBM Thomas J Watson Research Center, USA
Russell Rockne	City of Hope National Medical Center, USA
Maria Rodriguez Martinez	IBM, Zurich Research Laboratory, Switzerland
Venkata Pardhasaradhi Satagopam	University of Luxembourg and ELIXIR-Luxembourg, Luxembourg
Fabien Scalzo	University of California, Los Angeles, USA
Grant Schissler	University of Nevada, Reno, USA
Matthias Schlesner	German Cancer Research Center, Germany
Alexander Schoenhuth	Vrije Universiteit Amsterdam, The Netherlands
Russell Schwartz	Carnegie Mellon University, USA
Jacob Scott	Cleveland Clinic, USA
Amarda Shehu	George Mason University, USA
Feichen Shen	Mayo Clinic, USA
Yang Shen	Texas A&M University, USA
Xinghua Shi	The University of North Carolina at Charlotte, USA
Chi-Ren Shyu	University of Missouri-Columbia, USA
Sunghwan Sohn	Mayo Clinic, USA
Thomas Steinke	Zuse Institute Berlin, Germany
Angelique Stephanou	TIMC-IMAG, CNRS, France
Kyung Sung	University of California, Los Angeles, USA
Ahmad Tafti	Mayo Clinic, USA
Umit Topaloglu	Wake Forest School of Medicine, USA
Tamir Tuller	Tel Aviv University, Israel
Jack Tuszynski	University of Alberta, Canada
Alexander V. Alekseyenko	Medical University of South Carolina, USA
George Vasmatzis	Mayo Clinic, USA
Volpert Vitaly	Université Claude Bernard Lyon 1, France
Li-San Wang	University of Pennsylvania, USA
Yanshan Wang	Mayo Clinic, USA
Jeremy Warner	Vanderbilt University, USA
Mark Wass	University of Kent, UK
Lonnie Welch	Ohio University, USA
Kenneth Wertheim	University of Nebraska–Lincoln, USA
Matt Williams	Imperial College, UK
Dominik Wodarz	University of California, Irvine, USA
Yanji Xu	Nevada University Reno, USA
Habil Zare	University of Washington, USA
Meirav Zehavi	Ben-Gurion University, Israel
Alex Zelikovsky	Georgia State University, USA
Xuegong Zhang	Tsinghua University, China
Xiaoming Zheng	Central Michigan University, USA
Maryam Zolnoori	Mayo Clinic, USA

Additional Reviewers

Bhattacharya, Sutanu
Chen, Ken
Gertz, Mike
Levy, Doron
Manandhar, Dinesh
Nguyen, Hung
Tran, Duc

Sponsors

Keynote Talks

Breast Cancer Genomics: Tackling Complexity with Functional Genomics and Patient-Derived Organoids

Ron Bose

Washington University School of Medicine, USA

Abstract. Breast cancer is a heterogeneous disease with multiple molecular subtypes and three major clinical subtypes: hormone receptor positive, HER2 positive, and triple negative breast cancer. These three clinical subtypes are very important because they determine the drugs used for patient treatment. Cellular, molecular, and genomic understanding of breast cancer has resulted in new treatments for breast cancer. In 2019, the FDA approved an oral PIK3CA inhibitor for PIK3CA mutated, hormone receptor positive, Stage IV breast cancer and immunotherapy for triple negative, Stage IV breast cancer. Major challenges facing future research on breast cancer and other cancers are: (1) Interpreting genome sequencing results to better understand the effects and significance of new or under-characterized mutations, and (2) having platforms for rapid biological testing of hypotheses. I will provide examples of how my laboratory is trying to address both of these challenges.

High Dimensional Unsupervised Approaches for Dealing with Heterogeneity of Cell Populations and Proliferation of Algorithmic Tools

Yuval Kluger

Yale University School of Medicine, USA

Abstract. Revealing the clonal composition of a single tumor is essential for identifying cell subpopulations with metastatic potential in primary tumors or with resistance to therapies in metastatic tumors. Bulk sequencing technologies provide only an overview of the aggregate of numerous cells. We propose an evolutionary framework for deconvolving data from a bulk genome-wide experiment to infer the composition, abundance and evolutionary paths of the underlying cell subpopulations of a tumor. With advances in high throughput single cell techniques, we can in principle resolve these issues. However, these techniques introduce new challenges such as analyzing datasets of millions of cells, batch effects, missing values etc. We provide several algorithmic solutions for some of these challenges. Finally, a key challenge in bioinformatics is how to rank and combine the possibly conflicting predictions of several algorithms, of unknown reliability. We provide new mathematical insights of striking conceptual simplicity that explain mutual relationships between independent classifiers/algorithms. These insights enable the design of efficient, robust and reliable methods to rank the classifiers performances and construct improved predictions in the absence of ground truth.

Career Development Opportunities: The Government Can Help!

Susan Perkins

National Cancer Institute, USA

Abstract. The National Cancer Institute is committed to the training and support of the next generation of cancer researchers. The NCI funds training at extramural institutions across the nation, using funding mechanisms that include fellowships, career development awards, and institutional training grants. This session will provide a broad overview of this wide range of opportunities, with an emphasis on some new NCI programs for early-stage investigators, as well as some tips and tools for applicants.

Bringing Math into the Clinic:
Mathematical Oncology at City of Hope

Russell Rockne

City of Hope, USA

Abstract. In this keynote lecture, I will provide vignettes of applications of mathematical modeling aimed at use in the clinic within the Division of Mathematical Oncology at City of Hope. I will focus on the use of non-invasive imaging (MRI, PET) to calibrate and validate patient-specific mathematical models of cancer growth and response to therapy. Applications include primary brain tumors and breast cancer, with therapeutic applications including cell-based therapies, radiation therapy, and combination therapies. I will provide a forward-looking view of Mathematical Oncology at City of Hope and present clinical challenges that may be addressed with mathematical modeling.

Application of Functional Genomics to Oncology Practice: Opportunities, Successes, Failures and Barriers

Panos Anastasiadis and George Vasmatzis

Mayo Clinic, USA

Abstract. Radical improvement in cancer care can be accomplished by individualizing patient management via the application of genomics and functional model systems into clinical practice. Recent breakthroughs in immunotherapy (i.e. checkpoint inhibitors) and targeted therapies (i.e. NTRK inhibitors) have shown that therapy of advanced cancers might become agnostic to the organ of origin, arguing for a more individualized approach to patient care. Emerging genomics technologies, data integration and visualization platforms are powerful tools to determine the state of the individual's tumor and point to tailored treatments. Furthermore, an efficient combination of comprehensive genomics with 3D microcancer functional model systems can further refine treatment decisions. However, applying such disruptive technologies in clinical practice is not trivial. Regulatory, financial and clinical barriers will be discussed.

Recognition of Non-synonymous Somatic Mutations by Tumor Infiltrating Lymphocytes (TIL) in Metastatic Breast Cancer

Nikos Zacharakis

National Cancer Institute, USA

Abstract. Adoptive transfer of tumor infiltrating lymphocytes (TIL) can mediate long-term durable regression in patients with metastatic melanoma, a type of cancer which is characterized by a high number of mutated genes and pronounced lymphocytic infiltrate. Common epithelial cancers, including breast cancer, express far fewer somatic mutations than melanoma and the level of reactive TIL is limited. This pilot study investigated the ability to identify personalized non-synonymous somatic mutations in metastatic breast cancer lesions, to grow TIL that recognize the products of these mutations, and to adoptively transfer these TIL into patients with metastatic breast cancer, refractory to other treatments. Metastatic and primary tumor lesions from thirty one patients with breast cancer were studied in the Surgery branch, NCI, NIH and all of them were found to contain and express mutated genes (range: 4-1788 , median: 99). TIL recognized at least one (range: 1-10, median: 3) mutated product in 21 of 32 the patients (66%). Five evaluable patients with metastatic breast cancer, refractory to prior multiple lines of treatment, were treated with enriched mutation-reactive TIL in our ongoing pilot clinical trial. The immunogenicity of mutations in the majority of the patients with metastatic breast cancer can be the platform for an adoptive T cell transfer therapeutic approach targeting those mutated genes.

Inferring Tumor Evolution from Bulk and Single-cell Sequencing Data

Ben Raphael

Princeton University, USA

Abstract. Cancer is an evolutionary process driven by somatic mutations that accumulate in a population of cells. These mutations provide markers to infer the ancestral relationships between cells of a tumor, to describe populations of cells that are sensitive/resistant to treatment, or to study migrations between a primary tumor and distant metastases. However, such phylogenetic analyses are complicated by specific features of cancer sequencing data such as heterogeneous mixtures of cells present in bulk tumor sequencing data, undersampling in single-cell sequencing data, and large-scale genome rearrangements. In this talk, I will describe algorithms to address several problems in tumor evolution including: the inference of seeding patterns of metastases; the identification of copy number aberrations and whole-genome duplications in multi-sample sequencing data; and the integrated analysis of single-nucleotide mutations and copy number aberrations in single-cell sequencing data.

Contents

Special Track: Tumor Evolvability and Intra-tumor Heterogeneity

Phylogenies Derived from Matched Transcriptome Reveal the Evolution of Cell Populations and Temporal Order of Perturbed Pathways in Breast Cancer Brain Metastases

Yifeng Tao[1,2], Haoyun Lei[1,2], Adrian V. Lee[3], Jian Ma[1],
and Russell Schwartz[1,4](✉)

[1] Computational Biology Department, School of Computer Science,
Carnegie Mellon University, Pittsburgh, PA 15213, USA
[2] Joint Carnegie Mellon–University of Pittsburgh Ph.D. Program in Computational
Biology, Pittsburgh, PA 15213, USA
[3] Department of Pharmacology and Chemical Biology,
UPMC Hillman Cancer Center, Magee–Womens Research Institute,
University of Pittsburgh, Pittsburgh, PA 15213, USA
[4] Department of Biological Sciences, Carnegie Mellon University,
Pittsburgh, PA 15213, USA
russells@andrew.cmu.edu

Abstract. Metastasis is the mechanism by which cancer results in mortality and there are currently no reliable treatment options once it occurs, making the metastatic process a critical target for new diagnostics and therapeutics. Treating metastasis before it appears is challenging, however, in part because metastases may be quite distinct genomically from the primary tumors from which they presumably emerged. Phylogenetic studies of cancer development have suggested that changes in tumor genomics over stages of progression often results from shifts in the abundance of clonal cellular populations, as late stages of progression may derive from or select for clonal populations rare in the primary tumor. The present study develops computational methods to infer clonal heterogeneity and temporal dynamics across progression stages via deconvolution and clonal phylogeny reconstruction of pathway-level expression signatures in order to reconstruct how these processes might influence average changes in genomic signatures over progression. We show, via application to a study of gene expression in a collection of matched breast primary tumor and metastatic samples, that the method can infer coarse-grained substructure and stromal infiltration across the metastatic transition. The results suggest that genomic changes observed in metastasis, such as gain of the *ErbB* signaling pathway, are likely caused by early events in clonal evolution followed by expansion of minor clonal populations in metastasis (Algorithmic details, parameter settings, and proofs are provided in an Appendix with source code available at https://github.com/CMUSchwartzLab/BrM-Phylo).

© Springer Nature Switzerland AG 2019
G. Bebis et al. (Eds.): ISMCO 2019, LNCS 11826, pp. 3–24, 2019.
https://doi.org/10.1007/978-3-030-35210-3_1

Keywords: Breast cancer · Brain metastases · Phylogenetics · Deconvolution · Pathways · Gene modules

1 Introduction

Metastatic disease is the primary mechanism by which cancer results in patient mortality [6,7]. By the time metastases have appeared, there are generally no viable treatment options [14]. Successful treatment thus depends on treating not just the primary tumor but the seeds of metastasis that may linger after a seemingly successful remission. Identifying successful treatment options for metastasis is problematic, however, since the genomics of primary and metastatic tumors may be quite different even in single patients and metastatic cell populations may be poorly responsive to therapies effective on the primary tumor. Studies of cell-to-cell variation in cancers have revealed often substantial clonal heterogeneity in single tumors, with clonal populations sometimes dramatically shifting across progression stages [13]. Phylogenetic studies of clonal populations have been inconclusive on the typical evolutionary relationships between primary and metastatic tumors [35] and it remains a matter of debate whether changes in clonal composition occur primarily through ongoing clonal evolution, which results in novel clones with metastatic potential and resistance to therapy, or from selection on existing clonal heterogeneity already present at the time of first treatment [5,10]. The degree to which either answer is true has important implications for prospects for early detection or prophylactic treatment of metastasis.

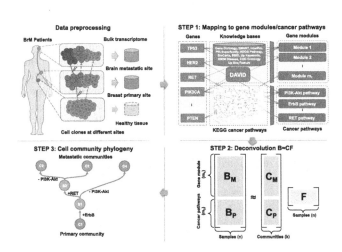

Fig. 1. The pipeline of BrM phylogenetics using matched bulk transcriptome.

Brain metastases (BrMs) occur in around 10%–30% of metastatic breast cancers cases [26]. Although recent advances in the treatment of metastatic breast cancer have been able to achieve long-term overall survival, there are limited treatment options for BrMs and clinical prognoses are still disappointing [41]. Recent work examining transcriptomic changes between paired primary and BrM samples has demonstrated dramatic changes in expression programs over metastasis, including changes in tumor subtype with important implications for treatment options and prognosis [31,38]. Some past research has sought to infer phylogenetic models to explain the development of brain metastases based on somatic genomic alterations [4,21]. Such methods are challenged in drawing robust conclusions about recurrent progression processes, though, by the high heterogeneity both within single tumors and across progression stages and patients. Changes in the activity of particular genetic pathways or modules may provide a more robust measure of frequent genomic alterations across cancers.

In the present work, we develop a strategy for tumor phylogenetics to explore how changes in clonal composition, via both novel molecular evolution and shifts in population dynamics of tumor clones and associated stroma, influence changes in expression programs across such progression stages. Our methods make use of multi-site bulk transcriptomic data to profile changes evident in gene expression programs between clones and progression stages. We break from past work in this domain in that we seek to study not clones *per se*, as is typical in tumor phylogenetics, but what we dub "cell communities": collections of clones or other stromal cell types that persist as a group with similar proportions across samples (Sect. 2.4). We accomplish this via a novel genomic deconvolution approach designed to make use of multiple samples both within and between patients [36] while improving robustness to inter- and intra-tumor heterogeneity by integrating deconvolution with pathway-based analyses of expression variation [30].

2 Methods

2.1 Overview

Cell populations evolve due to genomic perturbations that can result in changes in the activity of various functional pathways between clones. Our overall method for deriving coarse-grained portraits of cell community evolution at the pathway level is illustrated by Fig. 1. After the preprocessing of transcriptome data (Sect. 2.2), the overall workflow consists of three main steps: First, the bulk expression profiles are mapped into the gene module and pathway space using external knowledge bases to reduce redundancy, noise, sparsity, and to provide markers of expression variation for the subsequent analysis (Sect. 2.3). Second, a deconvolution step is implemented to resolve cell communities, i.e., coarse-grained mixtures of cell types presumed to represent an associated population of cancer clones and stromal cells, from the compressed pathway representation of samples (Sect. 2.4). Third, phylogenies of these cell communities are built based on the deconvolved communities as well as inferred ancestral (Steiner) communities to reconstruct likely trajectories of evolutionary progression by which cell

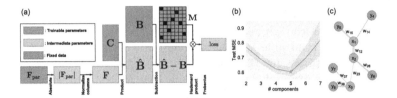

Fig. 2. Method details. (a) Neural network architecture of NND. (b) Test errors of NND using 20-fold CV. Errors in unit of mean square error (MSE). (c) Illustration of a phylogeny with five extant nodes and three Steiner nodes.

communities develop—through a combination of genetic mutations, expression changes, and changes in population distributions—as a tumor progresses from healthy tissue to primary and potentially metastatic tumor (Sect. 2.5).

2.2 Transcriptome Data Preprocessing

We applied a series of preprocessing methods, including quantile normalization [1], to the raw bulk RNA-Sequencing data of 44 matched primary breast and metastatic brain tumors from 22 patients [31,38]. See Appendix Sect. A1 for detailed protocols of data preprocessing.

2.3 Mapping to Gene Modules and Cancer Pathways

The mapping step compresses the high dimensional data and provides markers of cancer-related biological processes (Fig. 1 Step 1). **Gene Modules:** Genes in the same "gene modules" [8,37] are usually affected by a common set of somatic alterations [30], and therefore are co-expressed in cells. We mapped the protein-coding gene expressions into gene modules using the DAVID tool and external knowledge bases [17,18]. The z-scores of $m_1 = 109$ gene modules in all the $n = 44$ samples were represented as a matrix $\mathbf{B}_M \in \mathbb{R}^{m_1 \times n}$. **Cancer Pathways:** We extracted the 23 cancer-related pathways from the KEGG database [19]. An additional recurrently gained *RET pathway* was added [38]. See y-axis of Fig. 3d for the complete list of pathways. z-scores of $m_2 = 24$ cancer pathways were represented as $\mathbf{B}_P \in \mathbb{R}^{m_2 \times n}$. In summary, the raw gene expressions were compressed into the gene module/pathway representation $\mathbf{B} = \left[\mathbf{B}_M^\mathsf{T}, \mathbf{B}_P^\mathsf{T}\right]^\mathsf{T} \in \mathbb{R}^{m \times n}$. The gene module serves for accurately deconvolving and unmixing the cell communities, while the pathway serves as markers/probes and for interpretation purpose. We will refer to the compressed representation containing both gene modules and pathways as "pathway representation" for brevity if not specified. See Appendix Sect. A2 for further details of the mapping.

2.4 Deconvolution of Bulk Data

We applied a type of matrix factorization (MF) with constraints on the pathway-level expression signatures to deconvolve the communities/populations from primary and metastatic tumor samples (Fig. 1 Step 2) [22]. Note that common alternatives, such as principal components analysis (PCA) and non-negative matrix factorization (NMF) [23] are not amenable to this case, since PCA does not provide a feasible solution to the constrained problem, and the NMF does not apply to our mixture data which can be either positive or negative.

Cell Communities. We define a cell community to be a set of clones/clonal subpopulations and other cell types that propagate as a group during the evolution of a tumor. A community may be just a single subpopulation/clone, but is a more general concept in the sense that it usually involves multiple related clones and their associated stroma. For example, a set of immunogenic clones and the immune cells infiltrating them might collectively form a community that has a collective expression signature mixing signatures of the clones and associated immune cells, even if the individual cell types are not distinguishable from bulk expression data alone. While much work in this space has classically aimed to separate individual clones, or perhaps individual cell types more broadly defined, we note that deconvolution may be unable in principle to resolve distinct cell types if they are always co-located in similar proportions. Particularly when data is sparse and cell types are fit only approximately, as in the present work, a model with large complexity to deconvolve the fine-grained populations is prone to overfit. The community concept is intended in part to better describe the results we expect to achieve from the kind of data examined here and in part because identifying these communities is itself of interest in understanding how tumor cells coevolve with their stroma during progression and metastasis. Single-cell methods may provide an alternative, but are not amenable to preserved samples such as are needed when retrospectively studying primary tumors and metastases that may have been biopsied years apart.

Formulation of Deconvolution. With a matrix of bulk pathway values $\mathbf{B} \in \mathbb{R}^{m \times n}$, the deconvolution problem is to find a component matrix $\mathbf{C} = [\mathbf{C}_M^{\mathsf{T}}, \mathbf{C}_P^{\mathsf{T}}]^{\mathsf{T}} \in \mathbb{R}^{m \times k}$ that represents the inferred fundamental communities of tumors, and the corresponding set of mixture fractions $\mathbf{F} \in \mathbb{R}_+^{k \times n}$:

$$\min_{\mathbf{C},\mathbf{F}} \quad \|\mathbf{B} - \mathbf{CF}\|_{\mathrm{Fr}}^2, \tag{1}$$

$$\text{s.t.} \quad \mathbf{F}_{lj} \geq 0, \qquad\qquad l = 1, ..., k, \ j = 1, ..., n, \tag{2}$$

$$\sum_{l=1}^{k} \mathbf{F}_{lj} = 1, \qquad\qquad j = 1, ..., n, \tag{3}$$

where $\|\mathbf{X}\|_{\mathrm{Fr}}$ is the Frobenius norm. The column-wise normalization in Eq. (3) aims for recovering the biologically meaningful cell communities. In addition, they are equivalent to applying ℓ_1 regularizers and therefore enforce sparsity to the fraction matrix $\mathbf{F}$ (Appendix Fig. 5).

Neural Network Deconvolution. Although it is possible to build new algorithms for solving MF by adapting previous work [23], the additional but necessary constraints of Eqs. (2–3) make the optimization much harder to solve. For the problem Eqs. (1–3), one can prove that it does not generally guarantee convexity (Appendix Sect. A3.1). A slightly modified version of the algorithm to solve NMF with constraints may guarantee neither good fitting nor convergence [25]. Therefore, instead of revising existing MF algorithms, such as ALS-FunkSVD [3,12,22], we developed an algorithm which we call "neural network deconvolution" (NND) to solve the optimization problem using gradient descent. Specifically, the NND was implemented using backpropagation in the form of a neural network (Fig. 2a) with PyTorch package (https://pytorch.org/) [20,34], based on the revised constraints:

$$\min_{\mathbf{C},\mathbf{F}_{\mathrm{par}}} \quad \|\mathbf{B} - \mathbf{C}\mathbf{F}\|_{\mathrm{Fr}}^2 \,, \tag{4}$$

$$\mathrm{s.t.} \quad \mathbf{F} = \mathrm{cwn}\left(|\mathbf{F}_{\mathrm{par}}|\right), \tag{5}$$

where $|\mathbf{X}|$ applies element-wise absolute value, cwn$(\mathbf{X})$ is column-wise normalization, so that each column sums up to 1. The two operations of Eq. (5) naturally rephrase and remove the two constraints in Eqs. (2–3), and meanwhile fit the framework of neural networks. This implementation is easy to adapt to a wide range of optimization scenarios with various constraints, and has the flexibility of allowing for cross-validation to prevent overfitting.

Cross-Validation of NND. In order to find the best tradeoff between model complexity and overfitting, we used cross-validation (CV) with the "masking" method to choose the optimal number of components/communities $k = 5$ that has the smallest test error (Fig. 2b). Note that the actual number of cell populations is probably considerably larger than 5, and therefore each one of the five communities may contain multiple cell populations. Furthermore, it is likely that with sufficient numbers and precision of measurements, these communities could be more finely resolved into their constituent cell types. However $k = 5$ represents the largest hypothesis space of NND model that can be applied to the current dataset without severe overfitting.

See Appendix for details of NND, including architecture specifications (Sect. A3.2), hyperparameters (Sect. A3.3), evaluation of fitting ability (Sect. A3.4), sparsity of results (Sect. A3.5), and cross-validation implementation (Sect. A3.6).

2.5   Phylogeny of Inferred Cell Subcommunities and Pathway Inference of Steiner Nodes

We built "phylogenies" of cell subcommunities and estimated the pathway representation of unobserved (Steiner) nodes [27] inferred to be ancestral to them, with the goal of discovering critical communities that appear to be involved in the transition to metastasis and identifying the important changes of functions and expression pathways during this transition (Fig. 1 Step 3). Note that we are using the term "phylogeny" loosely here, as these trees are intended to capture evolution of populations of cells not just by accumulation of mutations from a single ancestral clone but also changes in community structure, for example due to generating or suppressing an immune response or migrating to a metastatic site. Although an abuse of terminology, we use the term phylogeny here to make clear the methodological similarity to more proper phylogenetic methods in wide use for analyzing mutational data in cancers [35].

Phylogeny of Communities. Given the pathway profiles of the extant communities at the time of collecting tumor samples $\mathbf{C} \in \mathbb{R}^{m \times k}$, a phylogeny of the k extant cell communities was built using the neighbor-joining (NJ) algorithm [29], which inferred a tree that contains k extant nodes/leaves, $k - 2$ unobserved Steiner nodes, and edges connecting two Steiner nodes or a Steiner node and an extant node. We estimated an evolutionary distance for any pair of two communities u, v as the input of NJ using the Euclidean distance between their pathway vectors $\|\mathbf{C}_{.u} - \mathbf{C}_{.v}\|_2$, similar to that in a prior work [30].

Inference of Pathways. Denote the phylogeny of cell subcommunities as $\mathcal{G} = (\mathcal{V}, \mathcal{E})$, and $\mathcal{V} = \mathcal{V}_S \cup \mathcal{V}_C$, where the indices of Steiner node $\mathcal{V}_S = \{1, 2, ..., k-2\}$, the indices of extant nodes $\mathcal{V}_C = \{k-1, k, ..., 2k-2\}$. For each edge $(u, v) \in \mathcal{E}$, where $1 \le u < v \le 2k - 2$, the first node of edge $u \le k - 2$ is always a Steiner node. The second node v can be either a Steiner node ($v \le k-2$) or extant node ($v \ge k - 1$). Denote the set of weights $\mathcal{W} = \{w_{uv} = 1/d_{uv} \mid (u, v) \in \mathcal{E}\}$ (inverse distance), where the edge length d_{uv} is the output of NJ. For each dimension i of the pathway vectors, we consider them independently and separately, so that each dimension of the Steiner nodes can be solved in the same way. Now let us consider the i-th dimension (and omit the subscript i for brevity) of extant nodes $\mathcal{V}_C$: $\mathbf{y} = [y_{k-1}, y_k, ..., y_{2k-2}]^\mathsf{T} = \mathbf{C}_i^\mathsf{T}$ and Steiner nodes $\mathcal{V}_S$: $\mathbf{x} = [x_1, x_2, ..., x_{k-2}]^\mathsf{T}$. Figure 2c illustrates a phylogeny where $k = 5$. The inference of the i-th element in the pathway vector of the Steiner nodes can be formulated as minimizing the following elastic potential energy $U(\mathbf{x}, \mathbf{y}; \mathcal{W})$:

$$\min_{\mathbf{x}} \quad U(\mathbf{x}, \mathbf{y}; \mathcal{W}) = \sum_{\substack{(u,v) \in \mathcal{E} \\ v \le k-2}} \frac{1}{2} w_{uv} (x_u - x_v)^2 + \sum_{\substack{(u,v) \in \mathcal{E} \\ v \ge k-1}} \frac{1}{2} w_{uv} (x_u - y_v)^2, \quad (6)$$

which can be further rephrased as a quadratic programming problem and solved easily. See Appendix Sect. A4 for the derivation and proof of this section.

3   Results

3.1   Gene Modules/Pathways Provide an Effective Representation

Gene expressions of samples were mapped into gene module and pathway space in order to reduce the noise of raw transcriptome data and reduce redundancy (Sect. 2.3). We verified that the pathway representation is effective in the sense that it captures distinguishing features of primary/metastatic sites and individual samples well and is able to identify recurrently gained or lost pathways.

Feature Space of the Pathway Representation. As one can see in Fig. 3a, the first principal component analysis (PCA) dimension of the pathway representation accounts for the difference between primary and metastatic samples,

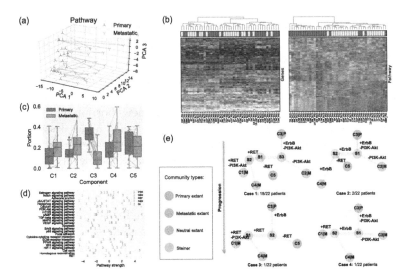

Fig. 3. Results and analysis. (a) First three pathway representation PCA dimensions of matched primary and metastatic samples. Matched samples are connected. (b) Hierarchical clustering of tumor samples based on raw gene expressions (left panel) and compressed gene module/pathway representation (right panel). Metastatic samples are shown in red rectangles and primary ones in yellow. (c) Portions and changes of the five communities in primary and metastatic sites. Each gray line connects the portions of a community in the primary site (blue node) and metastatic site (red node) from the same patient. (d) Pathway strengths across cell communities. (e) Phylogeny of cell subcommunities. (Color figure online)

while the second and third PCA dimensions mainly capture variability between patients. This observation suggests the feasibility of using the pathway representation to distinguish recurrent features of metastatic progression across patients despite heterogeneity between patients. To make a direct comparison of the noise and redundancy between the pathway and raw gene expression representations, we applied hierarchical clustering to the 44 samples using Ward's minimum variance method [39]. Two hierarchical trees were built based on the two different representations (Fig. 3b). The gene module/pathway features more effectively separate the primary and metastatic samples into distinct clusters (Fig. 3b right panel) than do the raw gene expression values (Fig. 3a left panel). This is consistent with the PCA results that the largest mode of variance in the pathway representation distinguishes primary from metastatic samples. We do notice that in a few cases, matched primary and metastatic samples from the same patient are neighbors with pathway-based clustering. For example, 29P_Pitt:29M_Pitt and 51P_Pitt:51M_Pitt are grouped in the same clades using the pathway representation, showing that in a minority of cases, features of individual patients dominate over primary vs. metastatic features. Following previous work [30], we quantified the ability of the hierarchical tree to group the samples of the same labels using four metrics. 1. MSD: Mean square distance of edges that connect nodes of the same label (primary vs. metastatic). 2. z_{MSD}: The labels of all nodes were shuffled and the MSD is recalculated for 1,000 times to get the mean μ_{MSD} and standard deviation σ_{MSD}, which were used to get the z-score of the current assignment $z_{\text{MSD}} = (\text{MSD} - \mu_{\text{MSD}})/\sigma_{\text{MSD}}$. 3. rMSD: The ratio of MSD of edges that connect same label nodes and MSD of edges that connect distinct label nodes. 4. z_{rMSD}: as with MSD, a z-score of rMSD was calculated by shuffling labels for 1,000 times. Intuitively, the smaller values the MSD, z_{MSD}, rMSD, and z_{rMSD} are, the better is the feature representation at grouping same label samples together. The shortest paths and distances between all pairs of nodes were calculated using the Floyd-Warshall algorithm [11,40]. All the edge length were considered as 1.0 to account for the different scales of pathway and gene representations. The pathway representation has significantly lower values for all four metrics (Table 1), indicating its strong grouping ability.

Recurrently Perturbed Cancer Pathways. We next identified differentially expressed pathways in the primary and metastatic tumors using bulk data $\mathbf{B}_P \in \mathbb{R}^{24 \times 44}$, prior to deconvolving cellular subcommunities. We conducted the Student's t-test followed by FDR correction on each of the 24 pathways. Eleven pathways are significantly different between the two sites (FDR < 0.05; Appendix Table 2). The signaling pathways related to neurotransmitter and calcium homeostasis (*cAMP, Calcium* [16]) are enriched in metastatic samples, which we can suggest may reflect stromal contamination by neural cells in the brain metastatic samples. We also observed recurrent gains in *ErbB* pathway, as indicated by the primary studies [31,38]. Three pathways related to immune activity are under-expressed in metastatic samples (*Cytokine-cytokine receptor interaction* [24], *JAK-STAT* [24], *Notch* [2]), consistent with the previous inference of reduced immune cell expression in metastases in general and brain metastasis most prominently [44]. We can suggest that this result similarly may reflect expression changes in infiltrating immune cells, due to the immunologically privileged environment of the brain, rather than expression changes in tumor cell populations. Five other signalling pathways (*Apoptosis* [42], *Wnt* [43], *Hedgehog* [15], *PI3K-Akt* [4], *TGF-beta* [28]) show reduction in metastatic samples and in each case, their loss or dysregulation has been reported to promote the tumor growth and brain metastasis. Note that the primary references for these data define pathways using co-expression pattern of genes [31,38], while our work uses external knowledge bases. Previous research also used somatic mutations or copy number variation to analyze perturbed genes [4,31], while we focus exclusively on the transcriptome. Despite large differences in data types and pathway definitions, our observations are consistent with the prior analysis, especially with respect to variation in the *HER2/ErbB2* and *PI3K-Akt* pathways.

3.2   Landscape of Deconvolved Cell Communities in Tumors

We unmixed the bulk data $\mathbf{B}$ into five components using NND (Sect. 2.4). The deconvolution enables us to produce at least a coarse-grained landscape of major cell communities $\mathbf{C}$ and their distributions in primary and metastatic tumors $\mathbf{F}$.

Table 1. Quantitative performance of hierarchical clustering.

Feature representation	MSD	rMSD	z_{MSD}	z_{rMSD}
Gene expression	99.62	0.93	−2.60	−2.57
Gene module/pathway	**86.23**	**0.66**	**−13.37**	**−11.42**

Community Distributions Across Samples F. The portions of the 5 components in all the 44 samples are represented as the mixture fraction matrix $\mathbf{F} \in \mathbb{R}^{5 \times 44}$ (Fig. 3c). A primary or metastatic community is one inferred to change proportions substantially (magnitude > 0.05) in the tumor samples after metastasis, or perhaps to be entirely novel to or extinct in the metastatic sample (denoted by a $|P$ or $|M$ suffix). Otherwise, the component is classified as a neutral community. Three components ($C1|M$, $C2|M$, $C4|M$) are classified as metastatic communities; one ($C3|P$) as primary; and one ($C5$) as neutral (Fig. 3c). Some components may be missing in both samples of some patients, e.g., $C1|M$, $C2|M$, $C5|M$ are absent in two, one, and one patient. We note that these five communities represent rough consensus clusters of cell populations inferred to occur frequently, but not universally, among the samples. Based on this rule, we can define four basic cases of patients in total. Twelve subcases can be found using a more detailed classification method based on the existence of communities in both primary and metastatic samples (Appendix Fig. 6).

Pathway Values of Communities C. We are especially interested in the pathway part $\mathbf{C}_P$ of the cell community inferences, since it serves as the marker and provides results easier to interpret. The pathway values of five subcommunities using $\mathbf{C}_P$ provides a much more fine-grained description of samples (Fig. 3d), compared with that in Sect. 3.1, which is only able to distinguish the differentially expressed pathways in bulk samples. As noted in Sect. 2.4, it is likely that true cellular heterogeneity is greater than the methods are able to discriminate and that communities inferred by our model may each conflate one or more distinct cell types and clones. We observe that the metastatic community $C4|M$ most prominently contributes to the enrichment for functions related to neurotransmitter and ion transport, since its strongest pathways (*cAMP*, *Calcium*) are greatly enriched relative to those of the other four communities. We might interpret this community as reflecting at least in part stromal contamination from neural cells specific to the metastatic site. $C4|M$ also contributes most to the gains of *ErbB* in brain samples. The metastatic subcommunity $C1|M$ is probably most closely related to the loss of immune response in metastatic samples as it has the lowest pathway values of *Notch*, *JAK-STAT*, and *Cytokine-cytokine receptor interaction*. This component might thus in part reflect the effect of relatively greater immune infiltration in the primary versus the metastatic site. $C1|M$ also has the lowest pathway values of *Apoptosis*, *Wnt*, and *Hedgehog*. The metastatic community $C2|M$ is most responsible for the loss of *PI3K-Akt* and *TGF-beta* pathways. We also note that although *RET* does not show up in the list of Table 2, it seems to be quite over-expressed in the metastatic communities $C1|M$ and $C4|M$ but not in the metastatic community $C2|M$.

3.3 Phylogenies of BrM Communities Reveal Common Temporal Order of Perturbed Pathways

We built phylogenies of cell communities and calculated the pathway representations of their Steiner nodes (Sect. 2.5). The phylogenies' topologies provide a way to infer a likely evolutionary history of cancer cell communities and thus their constitutive cell types, while the perturbed pathways along their edges suggest the temporal order of genomic alterations or changes in community composition.

Topologically Similar BrM Phylogenies. All five cell components do not appear in each BrM patient. We analyze the distribution of communities in each patient based on whether the community is inferred to be present in the patient (Appendix Fig. 6). There are four different cases in general (Fig. 3e). Case 1: all five communities are found in the patient (majority; 18/22 patients). Case 2: only $C1|M$ missing (minority; 2/22). Case 3: only $C2|M$ missing (minority; 1/22). Case 4: only $C5$ missing (minority; 1/22). Although not all communities exist in Case 2–4, the topologies are similar to that of Case 1 and can be seen as special cases of Case 1, representing some inferred common mechanisms of progression across all the BrM patients.

Common Temporal Order of Altered Cancer Pathways. After inferring the pathway values for Steiner nodes, the most perturbed pathways can also be found by subtracting the pathway vectors of nodes that share an edge. We focus on the top five gained or lost pathways along the evolutionary trajectories and the changes of magnitude larger than 1.0 (Appendix Tables 3, 4, 5 and 6). We further examine those perturbed cancer pathways that were specifically proposed in the study that generated the data examined here, as well as others that are clinically actionable [4,31,38], i.e., *ErbB*, *PI3K-Akt*, and *RET* (Fig. 3e). As one may see from Case 1, the primary community $C3|P$ first evolves to community $S3$ by gaining expressions in *ErbB* and losing functions in *PI3K-Akt*. Then, if it continues to lose *PI3K-Akt* activity, it will evolve into the metastatic community $C2|M$. If it gains in *RET* activity, it will instead evolve into metastatic communities $C1|M$ and $C4|M$. The perturbed pathways along the trajectories of Cases 2–4 are similar to those of Case 1, with minor differences. We therefore draw to the conclusion that the evolution of BrMs follows a specific and common order of pathway perturbations. Specifically, the gain of *ErbB* reproducibly happens before the loss of *PI3K-Akt* and the gain of *RET*. Different subsequently perturbed pathways lead to different metastatic tumor cell communities. These inferences are consistent with the hypothesis that at least some major changes in expression programs between primary and metastatic communities occur by selecting for heterogeneity present early in tumor development rather than solely deriving from novel functional changes immediately prior to or after metastasis.

4   Conclusions and Future Work

Cancer metastasis is usually a precursor to mortality with no successful treatment options. Better understanding mechanisms of metastasis provides a potential pathway to identify new diagnostics or therapeutic targets that might catch metastasis before it ensues, treat it prophylactically, or provide more effective treatment options once it occurs. The present work developed a computational approach intended to better reconstruct mechanisms of functional adaption from multisite RNA-seq data to help us understand at the level of cancer pathways the mechanisms by which progression frequently proceeds across a patient cohort. Our method compresses expression data into gene module/pathway representation using external knowledge bases, deconvolves the bulk data into putative cell communities where each community contains a set of associated cell types or subclones, and builds evolutionary trees of inferred communities with the goal of reconstructing how these communities evolve, adapt, and reconfigure their compositions across metastatic progression. We applied the pipeline to matched transcriptome data from 22 BrM patients and found that although there are slight differences of tumor communities across the cohort, most patients share a similar mechanism of tumor evolution at the pathway level. Specifically, the methods infer a fairly conserved mechanism of early gain of *ErbB* prior to metastasis, followed post-metastasis gain of *RET* or loss of *PI3K-Akt* resulting in intertumor heterogeneity between samples. Our methods provide a novel way of viewing the development of BrM with implications for basic research into metastatic processes and potential translational applications in finding markers or drug targets of metastasis-producing clones prior to the metastatic transition.

The results suggest several possible avenues for future development. In part, they suggest a need for better separating phylogenetically-related mixture components (i.e., distinct tumor cell clones) from unrelated infiltrating cell types (e.g., healthy stroma from the primary or metastatic site or infiltrating immune cells). The methods are likely finding only a small fraction of the true clonal heterogeneity of the tumors and stroma, and might benefit from algorithms capable of better resolution or from integration of multi-omics data (e.g., RNA-seq, DNA-seq, methylation) that might have complementary value in finer discrimination of cell types. Validation is challenging as we know of no data with known ground truth that models the kind of progression process studied here nor of other tools designed for modeling similar progression processes from expression data, leaving us reliant on validating based on consistency with prior research on brain metastasis [4,31,38]. Future work might compare to prior approaches for reconstruction of clonal evolution from expression data more generically [9,33,36] and seek replication on additional real or simulated expression data or artificial mixtures of different cell types [32] designed to mimic metastasis-like progression. The general approach might also have broader application than studying metastasis, for example in reconstructing mechanisms of other progression processes, such as pre-cancerous to cancerous, as well as to other tumor types or independent data sets. Finally, much remains to be done to exploit the translational potential of the method in better identifying diagnostic signatures and therapeutic targets.

Funding

This work was supported in part by a grant from the Mario Lemieux Foundation, U.S. N.I.H. award R21CA216452, Pennsylvania Department of Health award 4100070287, Breast Cancer Alliance, Susan G. Komen for the Cure, and by a fellowship to Y.T. from the Center for Machine Learning and Healthcare at Carnegie Mellon University. The Pennsylvania Department of Health specifically disclaims responsibility for any analyses, interpretations or conclusions.

References

1. Amaratunga, D., et al.: Analysis of data from viral DNA microchips. J. Am. Stat. Assoc. **96**(456), 1161–1170 (2001)
2. Aster, J.C., et al.: The varied roles of Notch in cancer. Ann. Rev. Pathol. **12**, 245–275 (2017)
3. Bell, R.M., et al.: Scalable collaborative filtering with jointly derived neighborhood interpolation weights. In: Seventh IEEE International Conference on Data Mining (ICDM 2007), pp. 43–52 (2007)
4. Brastianos, P.K., et al.: Genomic characterization of brain metastases reveals branched evolution and potential therapeutic targets. Cancer Discov. **5**(11), 1164–1177 (2015)
5. de Bruin, E.C., et al.: Spatial and temporal diversity in genomic instability processes defines lung cancer evolution. Science (New York N.Y.) **346**(6206), 251–256 (2014)
6. Chaffer, C.L., et al.: A perspective on cancer cell metastasis. Science **331**(6024), 1559–1564 (2011)
7. Chambers, A.F., et al.: Dissemination and growth of cancer cells in metastatic sites. Nat. Rev. Cancer **2**(8), 563–572 (2002)
8. Desmedt, C., et al.: Biological processes associated with breast cancer clinical outcome depend on the molecular subtypes. Clin. Cancer Res. **14**(16), 5158–5165 (2008)
9. Desper, R., et al.: Tumor classification using phylogenetic methods on expression data. J. Theor. Biol. **228**(4), 477–496 (2004)
10. Ding, L., et al.: Advances for studying clonal evolution in cancer. Cancer Lett. **340**(2), 212–219 (2013)
11. Floyd, R.W.: Algorithm 97: shortest path. Commun. ACM **5**(6), 344–348 (1962)
12. Funk, S.: Netflix update: try this at home (2006)
13. Greaves, M., et al.: Clonal evolution in cancer. Nature **481**(7381), 306–313 (2012)
14. Guan, X.: Cancer metastases: challenges and opportunities. Acta Pharm. Sinica B **5**(5), 402–418 (2015)
15. Gupta, S., et al.: Targeting the Hedgehog pathway in cancer. Ther. Adv. Med. Oncol. **2**(4), 237–250 (2010)
16. Hofer, A.M., et al.: Extracellular calcium and cAMP: second messengers as "third messengers"? Physiology **22**(5), 320–327 (2007)

17. Hosack, D.A., et al.: Identifying biological themes within lists of genes with EASE. Genome Biol. **4**(10), R70–R70 (2003)
18. Huang, D.W., et al.: Systematic and integrative analysis of large gene lists using DAVID bioinformatics resources. Nat. Protoc. **4**(1), 44–57 (2009)
19. Kanehisa, M., et al.: KEGG: kyoto encyclopedia of genes and genomes. Nucleic Acids Res. **28**(1), 27–30 (2000)
20. Kingma, D., et al.: Adam: a method for stochastic optimization. In: International Conference on Learning Representations, December 2014
21. Körber, V., et al.: Evolutionary trajectories of IDHWT glioblastomas reveal a common path of early tumorigenesis instigated years ahead of initial diagnosis. Cancer Cell **35**(4), 692–704 (2019)
22. Koren, Y., et al.: Matrix factorization techniques for recommender systems. Computer **42**(8), 30–37 (2009)
23. Lee, D.D., et al.: Algorithms for non-negative matrix factorization. In: Proceedings of the 13th International Conference on Neural Information Processing Systems, NIPS 2000, pp. 535–541. MIT Press, Cambridge (2000)
24. Lee, S., et al.: Cytokines in cancer immunotherapy. Cancers **3**(4), 3856–3893 (2011)
25. Lei, H., et al.: Tumor copy number deconvolution integrating bulk and single-cell sequencing data. In: Cowen, L.J. (ed.) RECOMB 2019. LNCS, vol. 11467, pp. 174–189. Springer, Cham (2019). https://doi.org/10.1007/978-3-030-17083-7_11
26. Lin, N.U., et al.: CNS metastases in breast cancer. J. Clin. Oncol. **22**(17), 3608–3617 (2004)
27. Lu, C.L., et al.: The full Steiner tree problem. Theor. Comput. Sci. **306**(1), 55–67 (2003)
28. Massagué, J.: TGFβ in cancer. Cell **134**(2), 215–230 (2008)
29. Nei, M., et al.: The neighbor-joining method: a new method for reconstructing phylogenetic trees. Mol. Biol. Evol. **4**(4), 406–425 (1987)
30. Park, Y., et al.: Network-based inference of cancer progression from microarray data. IEEE/ACM Trans. Comput. Biol. Bioinf. **6**(2), 200–212 (2009)
31. Priedigkeit, N., et al.: Intrinsic subtype switching and acquired ERBB2/HER2 amplifications and mutations in breast cancer brain metastases. JAMA Oncol. **3**(5), 666–671 (2017)
32. Qiu, P., et al.: Extracting a cellular hierarchy from high-dimensional cytometry data with SPADE. Nat. Biotechnol. **29**(10), 886–891 (2011)
33. Riester, M., et al.: A differentiation-based phylogeny of cancer subtypes. PLoS Comput. Biol. **6**(5), e1000777 (2010)
34. Rumelhart, D.E., et al.: Learning representations by back-propagating errors. Nature **323**, 533 (1986)
35. Schwartz, R., et al.: The evolution of tumour phylogenetics: principles and practice. Nat. Rev. Genet. **18**, 213 (2017)
36. Schwartz, R., et al.: Applying unmixing to gene expression data for tumor phylogeny inference. BMC Bioinform. **11**(1), 42 (2010)
37. Tao, Y., et al.: From genome to phenome: Predicting multiple cancer phenotypes based on somatic genomic alterations via the genomic impact transformer. In: Pacific Symposium on Biocomputing (2020)
38. Vareslija, D., et al.: Transcriptome characterization of matched primary breast and brain metastatic tumors to detect novel actionable targets. J. Natl Cancer Inst. **111**(4), 388–398 (2018)
39. Ward, J.H.: Hierarchical grouping to optimize an objective function. J. Am. Stat. Assoc. **58**(301), 236–244 (1963)

40. Warshall, S.: A theorem on boolean matrices. J. ACM **9**(1), 11–12 (1962)
41. Witzel, I., et al.: Breast cancer brain metastases: biology and new clinical perspectives. Breast Cancer Res. **18**(1), 8 (2016)
42. Wong, R.S.Y.: Apoptosis in cancer: from pathogenesis to treatment. J. Exp. Clinical Cancer Res. **30**(1), 87 (2011)
43. Zhan, T., et al.: Wnt signaling in cancer. Oncogene **36**, 1461 (2016)
44. Zhu, L., et al.: Metastatic breast cancers have reduced immune cell recruitment but harbor increased macrophages relative to their matched primary tumors. bioRxiv: 525071 (2019)

Appendix

A1   Transcriptome Data Preprocessing

We applied our methods to raw bulk RNA-Sequencing data of 44 matched primary breast and metastatic brain tumors from 22 patients (each patient gives two samples) [31,38], where six patients are from the Royal College of Surgeons (RCS) and sixteen patients from the University of Pittsburgh (Pitt). These data profiled the expression levels of approximately 60,000 transcripts. We removed the genes that are not expressed in any sample. We also considered only protein-coding genes in the present study. We conducted quantile normalization across samples using the geometric mean to remove possible artifacts from different experiment batches [1]. The top 2.5% and bottom 2.5% of expressions were clipped to further reduce noise. Finally, we transformed the resulting bulk gene expression values into the log space and mapped those for each gene to the interval $[0, 1]$ by a linear transformation.

A2   Mapping to Gene Modules and Cancer Pathways

The protein-coding gene expressions were mapped into both perturbed gene modules and cancer pathways, using the DAVID tool and external knowledge bases [18], as well as the cancer pathways in KEGG database [19]. This step compresses the high dimensional data and provides markers of cancer-related biological processes (Fig. 1 Step 1).

Gene Modules. Functionally similar genes are usually affected by a common set of somatic alterations [30] and therefore are co-expressed in the cells. These genes are believed to belong to the same "gene modules" [8,37]. Inspired by the idea of gene modules, we fed a subset of 3,000 most informative genes out of the approximately 20,000 genes that have the largest variances into the DAVID tool for functional annotation clustering using several databases [18]. DAVID maps each gene to one or more modules. We did not force the genes to be mapped into disjunct modules because a gene may be involved in several biological functions and therefore more than one gene module. We removed gene modules that were not enriched (fold enrichment < 1.0) and kept the remaining $m_1 = 109$ modules (and the corresponding annotated functions), where fold enrichment is defined as the EASE score of the current module to the geometric mean of EASE scores in all modules [17]. The gene module values of all the $n = 44$ samples were represented as a gene module matrix $\mathbf{B}_M \in \mathbb{R}^{m_1 \times n}$. The i-th gene module value in j-th sample, $\mathbf{B}_{M_{i,j}}$, was calculated by taking the sum of expressions of all the genes in the i-th module. Then $\mathbf{B}_M$ was rescaled row-wise by taking the z-scores across samples to compensate for the effect of variable module sizes.

Cancer Pathways. Although gene module representation is able to capture the variances across samples and reduce the redundancy of raw gene expressions, it has two disadvantages: First of all, lack of interpretability. Specifically, some annotations assigned by DAVID are not directly related to biological functions, and the annotations of different modules may substantially overlap. Secondly, the key perturbed cancer pathways or functions may not be always the ones that vary most across samples. For example, genes in cancer-related KEGG pathways (hsa05200) [19] are not especially enriched in the top 3,000 genes with the largest expression variances. To make better use of prior knowledge on cancer-relevant pathways, we supplemented the generic DAVID pathway sets with a KEGG "cancer pathway" representation of samples $\mathbf{B}_P \in \mathbb{R}^{m_2 \times n}$, where the number of cancer pathways $m_2 = 24$. The cancer-related pathways in the KEGG database are cleaner and easier to explain, more orthogonal to each other, and contain critical signaling pathways to cancer development. We extracted the 23 cancer-related pathways from the following 3 KEGG pathway sets: *Pathways in cancer* (hsa05200), *Breast cancer* (hsa05224), and *Glioma* (hsa05214). An additional cancer pathway *RET pathway* was added, since it was found to be recurrently gained in the prior research [38]. See y-axis of Fig. 3d for the complete list of 24 cancer pathways. We considered all the $\sim$20,000 protein-coding genes other than top 3,000 genes. The following mapping of cancer pathways and transformation to z-scores were similar to that we did to map the gene modules.

Until this step, the raw gene expressions of n samples were transformed into the compressed gene module/pathway representation of samples $\mathbf{B} = [\mathbf{B}_M^{\mathsf{T}}, \mathbf{B}_P^{\mathsf{T}}]^{\mathsf{T}} \in \mathbb{R}^{m \times n}$, where $m = m_1 + m_2$. The gene module representation $\mathbf{B}_M$ serves for accurately deconvolving and unmixing the cell communities, while the pathway representation $\mathbf{B}_P$ serves as markers/probes and for interpretation purposes.

A3   Deconvolution of Bulk Data

A3.1   Non-convexity of Deconvolution Problem

Theorem 1. *The deconvolution problem Eqs. (1–3) below is not convex:*

$$\min_{C,F} \quad f(C, F) = \|B - CF\|_{\mathrm{Fr}}^2, \tag{7}$$

$$s.t. \quad F_{lj} \geq 0, \qquad\qquad l = 1, ..., k, \ j = 1, ..., n, \tag{8}$$

$$\sum_{l=1}^{k} F_{lj} = 1, \qquad\qquad j = 1, ..., n. \tag{9}$$

Proof. If the problem is convex, we should have: $\forall \lambda \in (0, 1)$, and $\forall \mathbf{C}_x, \mathbf{C}_y, \mathbf{F}_x, \mathbf{F}_y$ in the feasible domain, the following inequality always holds:

$$\lambda f(\mathbf{C}_x, \mathbf{F}_x) + (1 - \lambda)f(\mathbf{C}_y, \mathbf{F}_y) \geq f(\lambda\mathbf{C}_x + (1 - \lambda)\mathbf{C}_y, \lambda\mathbf{F}_x + (1 - \lambda)\mathbf{F}_y). \tag{10}$$

However, for the following setting:

$$\mathbf{B} = \begin{bmatrix} -1.38 & 0.92 \\ 1.03 & -0.15 \end{bmatrix}, \tag{11}$$

$$\mathbf{C}_x = \begin{bmatrix} -1.74 & 2.21 \\ 1.00 & -3.97 \end{bmatrix}, \qquad \mathbf{C}_y = \begin{bmatrix} 1.03 & -0.46 \\ -3.13 & 0.16 \end{bmatrix}, \tag{12}$$

$$\mathbf{F}_x = \begin{bmatrix} 0.83 & 0.32 \\ 0.17 & 0.68 \end{bmatrix}, \qquad \mathbf{F}_y = \begin{bmatrix} 0.09 & 0.34 \\ 0.91 & 0.66 \end{bmatrix}, \tag{13}$$

and $\lambda = 0.5$, we have

$$\lambda f(\mathbf{C}_x, \mathbf{F}_x) + (1 - \lambda)f(\mathbf{C}_y, \mathbf{F}_y) = 4.86 < 11.74 = f(\lambda\mathbf{C}_x + (1 - \lambda)\mathbf{C}_y, \lambda\mathbf{F}_x + (1 - \lambda)\mathbf{F}_y). \tag{14}$$

This is contradictory to Eq. (10). □

A3.2   Architecture Specifications of NND

In the NND architecture, $|\mathbf{X}|$ applies element-wise absolute value, cwn $(\mathbf{X})$ column-wisely normalizes $\mathbf{X}$, so that each column of the output sums up to 1. The two operations of Eq. (5) naturally rephrase and remove the two constraints in Eqs. (2–3), and meanwhile fit the framework of neural networks. An alternative to the absolute value operation $|\mathbf{X}|$ might be rectified linear unit ReLU$(\mathbf{X}) = \max(\mathbf{0}, \mathbf{X})$. However, this activation function is unstable and leads to inferior performance in our case, since $\mathbf{X}_{lj}$ will be fixed to zero once it becomes negative and will lose the chance to get updated in the following iterations. One may also want to replace the column-wise normalization cwn $(\mathbf{X})$ with softmax operation softmax$(\mathbf{X})$. However, the nonlinearity introduced by softmax actually changes the original optimization problem Eqs. (1–3) and the fitted $\mathbf{F}$ is therefore not sparse.

A3.3   Hyperparameters of NND

We used an Adam optimizer with default momentum parameters and learning rate of 1×10^{-5} [20]. The mini-batch technique is not required since the data size in our application is small enough not to require it ($\mathbf{B} \in \mathbb{R}^{m \times n}$, $m = 133$, $n = 44$). The training was run until convergence, when the relative decrease of training loss is smaller than $\epsilon = 1 \times 10^{-10}$ every 20,000 iterations.

A3.4   Fitting Ability of NND

One might be suspicious whether the neural network fits precisely in practice, since it is based on a simple gradient descent optimization. To validate the fitting ability of NND, we plotted the PCA of original samples $\mathbf{B}$ and the fitted samples $\hat{\mathbf{B}} = \mathbf{CF}$ (Fig. 4). One can easily see that NND provides good model fits to the data.

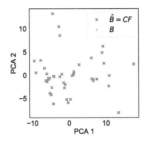

Fig. 4. PCA of pathway representation $\mathbf{B}$ and nnMF fitted $\hat{\mathbf{B}}$. Each dot represents the pathway values of a sample $\mathbf{B}_{.j}$ or fitted $\hat{\mathbf{B}}_{.j}$. The first two PCA dimensions of original data and fitted data are almost in the same positions, which indicates that NND is able to fit precisely in our application. The number of components is set to be $k = 5$ here.

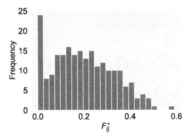

Fig. 5. Distribution of elements in fraction matrix $\mathbf{F}^{\star}$. Since each column of $\mathbf{F}$ is forced to sum up to be one, a Laplacian prior is applied to the elements of matrix $\mathbf{F}$. This leads to the sparsity of $\mathbf{F}^{\star}$: 24 out of its 220 elements ($k \times n = 5 \times 44$) are zeros (threshold set to 2.5×10^{-2}).

A3.5   Sparsity of NND Results

See Fig. 5 for distribution of fraction matrix in NND deconvolution results.

A3.6   Cross-Validation of NND

In each fold of the CV, we used $\hat{\mathbf{B}} = \mathbf{CF}$ to only fit some randomly selected elements of $\mathbf{B}$, and then the test error was calculated using the other elements of $\mathbf{B}$. This was implemented by introducing two additional mask matrices $\mathbf{M}_{\text{train}}, \mathbf{M}_{\text{test}} \in \{0,1\}^{m \times n}$, which are in the same shape of $\mathbf{B}$, and $\mathbf{M}_{\text{train}} + \mathbf{M}_{\text{test}} = \mathbf{1}^{m \times n}$. During the training time, with the same constraints in Eq. (5), the optimization goal is:

$$\min_{\mathbf{C}, \mathbf{F}_{\text{par}}} \|\mathbf{M}_{\text{train}} \odot (\mathbf{B} - \mathbf{CF})\|_{\text{Fr}}^2 \qquad (15)$$

where $\mathbf{X} \odot \mathbf{Y}$ is the Hadamard (element-wise) product. At the time of evaluation, given optimized $\mathbf{C}^\star$, $\mathbf{F}_{\text{par}}^\star$, and therefore optimized $\mathbf{F}^\star = \text{cwn}\left(|\mathbf{F}_{\text{par}}^\star|\right)$ for the optimization problem during training, the test error was calculated on the test set: $\|\mathbf{M}_{\text{test}} \odot (\mathbf{B} - \mathbf{C}^\star \mathbf{F}^\star)\|_{\text{Fr}}^2$. We used 20-fold cross-validation on the NND, so in each fold 95% positions of $\mathbf{M}_{\text{train}}$ and 5% positions of $\mathbf{M}_{\text{test}}$ were 1s.

A4   Derivation of Quadractic Programming, $\mathbf{P}(\mathcal{W})$, and $\mathbf{q}(\mathcal{W}, \mathbf{c})$

Recall Sect. 2.5, for the phylogeny $\mathcal{G} = (\mathcal{V}, \mathcal{E})$, the Steiner nodes are indexed as $\mathcal{V}_S = \{1, 2, ..., k-2\}$ ($|\mathcal{V}_S| = k - 2$), the extant nodes are indexed as $\mathcal{V}_C = \{k-1, k, ..., 2k-2\}$ ($|\mathcal{V}_C| = k$). The i-th pathway values of Steiner nodes are denoted as $\mathbf{x} = [x_1, x_2, ..., x_{k-2}]^\mathsf{T} \in \mathbb{R}^{k-2}$, and values of extant nodes as $\mathbf{y} = [y_{k-1}, y_k, ..., y_{2k-2}]^\mathsf{T} \in \mathbb{R}^k$. Since we consider each pathway dimension separately here, the subscript i for $\mathbf{x}$ and $\mathbf{y}$ is omitted for brevity. The weight of edge $(u, v) \in \mathcal{E}$ connecting nodes u and v is w_{uv} ($1 \le u < v \le 2k - 2$). Denote $\mathcal{W} = \{w_{uv} \mid (u, v) \in \mathcal{E}\}$. The inference of the i-th element in the pathway vector of the Steiner nodes can be formulated as minimizing the elastic potential energy $U(\mathbf{x}, \mathbf{y}; \mathcal{W})$ shown below:

$$\min_{\mathbf{x}} \quad U(\mathbf{x}, \mathbf{y}; \mathcal{W}) = \sum_{\substack{(u,v) \in \mathcal{E} \\ v \le k-2}} \frac{1}{2} w_{uv}(x_u - x_v)^2 + \sum_{\substack{(u,v) \in \mathcal{E} \\ v \ge k-1}} \frac{1}{2} w_{uv}(x_u - y_v)^2, \qquad (16)$$

Theorem 2. *Equation (16) can be further rephrased as a quadratic programming problem:*

$$\min_{x} \quad \frac{1}{2} \mathbf{x}^\mathsf{T} \mathbf{P}(\mathcal{W}) \mathbf{x} + \mathbf{q}(\mathcal{W}, \mathbf{y})^\mathsf{T} \mathbf{x}, \qquad (17)$$

where $\mathbf{P}(\mathcal{W})$ is a function that takes as input edge weights $\mathcal{W}$ and outputs a matrix $\mathbf{P} \in \mathbb{R}^{(k-2) \times (k-2)}$, $\mathbf{q}(\mathcal{W}, \mathbf{y})$ is a function that takes as input edge weights $\mathcal{W}$ and vector $\mathbf{y}$ and outputs a vector $\mathbf{q} \in \mathbb{R}^{k-2}$.

Proof. Based on Eq. (16), $U(\mathbf{x}, \mathbf{y}; \mathcal{W}) \geq 0$. Each term inside the first summation $(v \leq k - 2)$ can be written as:

$$\frac{1}{2} w_{uv}(x_u - x_v)^2 = \frac{1}{2}\mathbf{x}^\mathsf{T}\mathbf{P}(w_{uv})\mathbf{x}, \tag{18}$$

where

$$\mathbf{P}(w_{uv}) = \begin{array}{c} \\ \text{u-th row} \\ \\ \text{v-th row} \\ \\ \end{array} \overset{\displaystyle \overset{\text{u-th col}\quad\text{v-th col}}{}}{\begin{bmatrix} 0 & 0 & 0 & 0 & 0 \\ 0 & w_{uv} & 0 & -w_{uv} & 0 \\ 0 & 0 & 0 & 0 & 0 \\ 0 & -w_{uv} & 0 & w_{uv} & 0 \\ 0 & 0 & 0 & 0 & 0 \end{bmatrix}}. \tag{19}$$

Each term $(v \geq k - 1)$ inside the second summation can be rephrased as:

$$\frac{1}{2} w_{uv}(x_u - y_v)^2 = \frac{1}{2}\mathbf{x}^\mathsf{T}\mathbf{P}(w_{uv})\mathbf{x} + \mathbf{q}(w_{uv}, y_v)^\mathsf{T}\mathbf{x} + C(w_{uv}, y_v), \tag{20}$$

where

$$\mathbf{P}(w_{uv}) = \begin{array}{c} \\ \text{u-th row} \\ \\ \\ \\ \end{array} \overset{\displaystyle \overset{\text{u-th col}}{}}{\begin{bmatrix} 0 & 0 & 0\,0\,0 \\ 0 & w_{uv} & 0\,0\,0 \\ 0 & 0 & 0\,0\,0 \\ 0 & 0 & 0\,0\,0 \\ 0 & 0 & 0\,0\,0 \end{bmatrix}}, \quad \mathbf{q}(w_{uv}, y_v) = \begin{array}{c} \\ \text{u-th row} \\ \\ \\ \end{array} \begin{bmatrix} 0 \\ -w_{uv}y_v \\ 0 \\ 0 \\ 0 \end{bmatrix}, \tag{21}$$

and $C(w_{uv}, y_v) = \frac{1}{2}w_{uv}y_v^2$ is independent of $\mathbf{x}$. Therefore the optimization in Eq. (16) can be calculated and written as below:

$$\min_{\mathbf{x}} \quad \sum_{\substack{(u,v)\in\mathcal{E} \\ v\leq k-2}} \frac{1}{2}\mathbf{x}^\mathsf{T}\mathbf{P}(w_{uv})\mathbf{x} + \sum_{\substack{(u,v)\in\mathcal{E} \\ v\geq k-1}} \left(\frac{1}{2}\mathbf{x}^\mathsf{T}\mathbf{P}(w_{uv})\mathbf{x} + \mathbf{q}(w_{uv}, y_v)^\mathsf{T}\mathbf{x}\right), \tag{22}$$

$$\Leftrightarrow \min_{\mathbf{x}} \quad \frac{1}{2}\mathbf{x}^\mathsf{T}\left(\sum_{\substack{(u,v)\in\mathcal{E} \\ v\leq k-2}} \mathbf{P}(w_{uv}) + \sum_{\substack{(u,v)\in\mathcal{E} \\ v\geq k-1}} \mathbf{P}(w_{uv})\right)\mathbf{x} + \sum_{\substack{(u,v)\in\mathcal{E} \\ v\geq k-1}} \mathbf{q}(w_{uv}, y_v)^\mathsf{T}\mathbf{x},$$
$$\tag{23}$$

$$\Leftrightarrow \min_{\mathbf{x}} \quad \frac{1}{2}\mathbf{x}^\mathsf{T}\mathbf{P}(\mathcal{W})\mathbf{x} + \mathbf{q}(\mathcal{W}, \mathbf{y})^\mathsf{T}\mathbf{x}. \tag{24}$$

$\square$

Remark 1. The optimal $\mathbf{x}^\star$ of the Eq. (16), or the solution to the quadratic programming problem Eq. (17) can be solved by setting the gradient to be $\mathbf{0}$:

$$\mathbf{P}(\mathcal{W})\mathbf{x}^\star + \mathbf{q}(\mathcal{W}, \mathbf{y}) = \mathbf{0}. \tag{25}$$

Therefore,

$$\mathbf{x}^\star = -\mathbf{P}(\mathcal{W})^{-1}\mathbf{q}(\mathcal{W}, \mathbf{y}). \tag{26}$$

Remark 2. Based on the proof, we can derive how to calculate the matrix $\mathbf{P}(\mathcal{W})$ and vector $\mathbf{q}(\mathcal{W}, \mathbf{y})$.

Initialize the matrix and vector with zeros:

$$\mathbf{P} \leftarrow 0^{(k-2)\times(k-2)}, \quad \mathbf{q} \leftarrow 0^{k-2}. \tag{27}$$

For each edge $(u, v) \in \mathcal{E}$ with weight w_{uv}, there are two possibilities of nodes u and v: First, if both of them are Steiner nodes ($u \leq k-2$, $v \leq k-2$), we update $\mathbf{P}$ and keep $\mathbf{q}$ the same:

$$\mathbf{P}_{uu} \leftarrow \mathbf{P}_{uu} + w_{uv}, \ \mathbf{P}_{vv} \leftarrow \mathbf{P}_{vv} + w_{uv}, \ \mathbf{P}_{uv} \leftarrow \mathbf{P}_{uv} - w_{uv}, \ \mathbf{P}_{vu} \leftarrow \mathbf{P}_{vu} - w_{uv}. \tag{28}$$

Second, if u is Steiner node and v is an extant node ($u \leq k-2$, $v \geq k-1$), we update both $\mathbf{P}$ and $\mathbf{q}$:

$$\mathbf{P}_{uu} \leftarrow \mathbf{P}_{uu} + w_{uv}, \quad \mathbf{q}_u \leftarrow \mathbf{q}_u - y_v \cdot w_{uv}. \tag{29}$$

We apply the same procedure to all dimension of pathways $i = 1, 2, ..., m$ to get the full pathway values for each Steiner node.

A5   Differentially Expressed Cancer Pathways

Table 2 provides a list of the identified differentially expressed cancer pathways.

Table 2. Differentially expressed cancer pathways between primary and metastatic samples (FDR < 0.05).

Gain/Loss after metastasis	Differentially expressed pathways	FDR
Relative gain	cAMP signaling pathway	6.88e-03
Relative gain	ErbB signaling pathway	2.09e-02
Relative gain	Calcium signaling pathway	4.39e-02
Relative loss	Cytokine-cytokine receptor interaction	4.37e-06
Relative loss	Apoptosis	8.53e-04
Relative loss	JAK-STAT signaling pathway	8.53e-04
Relative loss	Wnt signaling pathway	3.97e-03
Relative loss	Hedgehog signaling pathway	4.50e-03
Relative loss	PI3K-Akt signaling pathway	1.35e-02
Relative loss	TGF-beta signaling pathway	4.56e-02
Relative loss	Notch signaling pathway	4.56e-02

A6   Portions of Cell Communities in BrM Patients

Figure 6 shows the inferred cell community portions across the BrM samples. The figure displays, for each patient, the proportion of each community in the primary and the metastatic sample.

A7   Perturbed Cancer Pathways Along Phylogenies

Tables 3, 4, 5 and 6 provide a full list of perturbed pathways across the phylogenies for Case 1, 2, 3, and 4 in Fig. 3e.

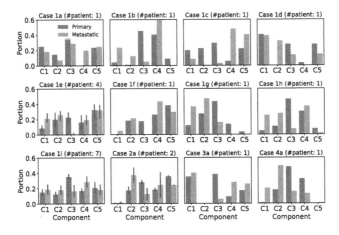

Fig. 6. Classification of BrM patients based on the consisted cell subcommunities in matched samples. There are 12 subcases of the 4 cases mentioned in Sect. 3.2. Specifically, there are 9 specific cases (Case 1a-i) in Case 1. Most patients (7) have all the five cell communities in both primary and metastatic samples (Case 1i). A few patients (4) have all communities in metastasis samples and all clones but community $C3|P$ in primary samples. The element $\mathbf{F}_{lj}$ is taken as 0 when it is smaller than a threshold 2.5×10^{-2}, and therefore the l-th community is missing in the j-th sample.

Table 3. Perturbed pathways during the evolution of cell communities in primary and metastatic tumors (Fig. 3e Case 1). The top five perturbed pathways whose gain or loss greater than 1.0 along each edge of phylogeny are shown. Clinically actionable perturbed cancer pathways during metastasis are shown in boldface, i.e., *ErbB*, *RET*, and *PI3K-Akt* [4,31,38].

Trajectory	Gain	Perturbed Pathways	Loss	Perturbed pathways
$C3\|P \to S3$	+2.83	Homologous recombination	−3.76	Hedgehog signaling pathway
	+2.41	Cell cycle	−3.45	Cytokine-cytokine receptor interaction
	+1.86	**ErbB signaling pathway**	−3.39	**PI3K-Akt signaling pathway**
	+1.10	cAMP signaling pathway	−3.15	TGF-beta signaling pathway
			−3.14	JAK-STAT signaling pathway
$S3 \to S1$	< 1.0	∅	< 1.0	∅
$S1 \to S2$	+1.36	cAMP signaling pathway	−1.28	JAK-STAT signaling pathway
	+1.18	**RET**	−1.22	Apoptosis
			−1.21	Cytokine-cytokine receptor interaction
			−1.12	Wnt signaling pathway
			−1.04	Notch signaling pathway
$S2 \to C1\|M$	+1.90	**RET**	−3.25	Wnt signaling pathway
	+1.59	PPAR signaling pathway	−3.11	JAK-STAT signaling pathway
			−2.77	Notch signaling pathway
			−2.48	Hedgehog signaling pathway
			−2.18	**PI3K-Akt signaling pathway**
$S2 \to C4\|M$	+4.48	Calcium signaling pathway	−3.06	p53 signaling pathway
	+4.17	cAMP signaling pathway	−2.74	Cell cycle
	+3.83	MAPK signaling pathway	−2.21	Homologous recombination
	+3.35	ECM-receptor interaction	−1.40	Apoptosis
	+3.20	Focal adhesion	−1.33	Cytokine-cytokine receptor interaction
$S1 \to C5$	+3.91	Cell cycle	−3.00	**RET**
	+3.17	p53 signaling pathway	−1.58	MAPK signaling pathway
	+2.85	Adherens junction	−1.41	cAMP signaling pathway
	+2.76	Cytokine-cytokine receptor interaction		
	+2.68	Wnt signaling pathway		
$S3 \to C2\|M$	+1.39	Homologous recombination	−3.65	TGF-beta signaling pathway
			−3.61	**PI3K-Akt signaling pathway**
			−3.34	ECM-receptor interaction
			−3.20	Focal adhesion
			−2.60	PPAR signaling pathway

Table 4. Perturbed pathways during the evolution of cell communities in primary and metastatic tumors (Fig. 3e Case 2). The top five perturbed pathways whose gain or loss greater than 1.0 along each edge of phylogeny are shown.

Trajectory	Gain	Perturbed Pathways	Loss	Perturbed pathways
$C3|P \to S1$	+2.83	Homologous recombination	−3.22	Hedgehog signaling pathway
	+2.47	Cell cycle	−3.10	TGF-beta signaling pathway
	+1.81	**ErbB signaling pathway**	−3.08	Cytokine-cytokine receptor interaction
	+1.02	cAMP signaling pathway	−2.93	**PI3K-Akt signaling pathway**
			−2.64	PPAR signaling pathway
$S1 \to S2$	+1.08	ECM-receptor interaction		
	+1.08	**ErbB signaling pathway**		
$S2 \to C4|M$	+5.51	cAMP signaling pathway	−3.97	Cell cycle
	+5.12	Calcium signaling pathway	−3.83	p53 signaling pathway
	+4.45	MAPK signaling pathway	−3.20	Apoptosis
	+3.37	ECM-receptor interaction	−3.15	Cytokine-cytokine receptor interaction
	+3.08	**ErbB signaling pathway**	−3.00	Homologous recombination
$S2 \to C5$	+3.68	Cell cycle	−2.25	**RET**
	+3.18	p53 signaling pathway	−1.81	MAPK signaling pathway
	+2.50	Homologous recombination	−1.43	cAMP signaling pathway
	+2.16	Adherens junction	−1.24	Hedgehog signaling pathway
	+2.15	Cytokine-cytokine receptor interaction	−1.13	Calcium signaling pathway
$S1 \to C2|M$	+1.39	Homologous recombination	−4.06	**PI3K-Akt signaling pathway**
			−3.70	TGF-beta signaling pathway
			−3.55	Focal adhesion
			−3.52	ECM-receptor interaction
			−2.87	Adherens junction

Table 5. Perturbed pathways during the evolution of cell communities in primary and metastatic tumors (Fig. 3e Case 3). The top five perturbed pathways whose gain or loss greater than 1.0 along each edge of phylogeny are shown.

Trajectory	Gain	Perturbed Pathways	Loss	Perturbed pathways
$C3\|P \to S2$	+3.10	Cell cycle	−3.51	Hedgehog signaling pathway
	+3.10	**ErbB signaling pathway**	−2.41	Notch signaling pathway
	+2.93	Homologous recombination	−2.39	Cytokine-cytokine receptor interaction
	+1.70	cAMP signaling pathway	−2.34	JAK-STAT signaling pathway
	+1.66	HIF-1 signaling pathway	−2.07	Apoptosis
$S2 \to S1$	+1.62	cAMP signaling pathway	−2.02	Cytokine-cytokine receptor interaction
	+1.54	**RET**	−1.98	JAK-STAT signaling pathway
	+1.14	Calcium signaling pathway	−1.91	Apoptosis
			−1.75	Wnt signaling pathway
			−1.32	Cell cycle
$S1 \to C1\|M$	+1.85	**RET**	−3.52	Wnt signaling pathway
	+1.19	PPAR signaling pathway	−3.38	JAK-STAT signaling pathway
			−2.78	**PI3K-Akt signaling pathway**
			−2.76	Hedgehog signaling pathway
			−2.68	Notch signaling pathway
$S1 \to C4\|M$	+4.20	Calcium signaling pathway	−3.18	p53 signaling pathway
	+3.89	cAMP signaling pathway	−2.65	Cell cycle
	+3.40	MAPK signaling pathway	−1.99	Homologous recombination
	+2.76	Hedgehog signaling pathway	−1.64	Cytokine-cytokine receptor interaction
	+2.72	ECM-receptor interaction	−1.61	Apoptosis
$S2 \to C5$	+3.67	Cell cycle	−2.69	**RET**
	+2.76	Homologous recombination	−2.08	MAPK signaling pathway
	+2.56	p53 signaling pathway	−1.59	PPAR signaling pathway
	+1.85	mTOR signaling pathway	−1.43	cAMP signaling pathway
	+1.79	Adherens junction	−1.02	Hedgehog signaling pathway

Table 6. Perturbed pathways during the evolution of cell communities in primary and metastatic tumors (Fig. 3e Case 4). The top five perturbed pathways whose gain or loss greater than 1.0 along each edge of phylogeny are shown.

Trajectory	Gain	Perturbed Pathways	Loss	Perturbed pathways
$C3\|P \to S1$	+2.38	Homologous recombination	−4.49	Cytokine-cytokine receptor interaction
	+1.56	**ErbB signaling pathway**	−4.23	**PI3K-Akt signaling pathway**
	+1.54	Cell cycle	−4.10	JAK-STAT signaling pathway
	+1.41	cAMP signaling pathway	−3.97	Hedgehog signaling pathway
			−3.74	Apoptosis
$S1 \to S2$	+1.89	cAMP signaling pathway	−1.66	Notch signaling pathway
	+1.69	**ErbB signaling pathway**	−1.27	JAK-STAT signaling pathway
	+1.47	HIF-1 signaling pathway	−1.14	Apoptosis
	+1.47	ECM-receptor interaction	−1.01	Cytokine-cytokine receptor interaction
	+1.43	Calcium signaling pathway		
$S2 \to C1\|M$	+1.43	PPAR signaling pathway	−2.53	Notch signaling pathway
	+1.19	**RET**	−2.44	Wnt signaling pathway
	+1.09	p53 signaling pathway	−2.35	Hedgehog signaling pathway
			−2.32	JAK-STAT signaling pathway
			−1.66	VEGF signaling pathway
$S2 \to C4\|M$	+4.40	Calcium signaling pathway	−2.37	p53 signaling pathway
	+3.91	cAMP signaling pathway	−1.93	Cell cycle
	+3.81	ECM-receptor interaction	−1.74	Homologous recombination
	+3.64	MAPK signaling pathway		
	+3.62	Focal adhesion		
$S1 \to C2\|M$	+1.84	Homologous recombination	−3.07	TGF-beta signaling pathway
	+1.39	Cell cycle	−2.77	**PI3K-Akt signaling pathway**
			−2.69	ECM-receptor interaction
			−2.59	Focal adhesion
			−2.58	PPAR signaling pathway

Modeling the Evolution of Ploidy in a Resource Restricted Environment

Gregory Kimmel[1], Jill Barnholtz-Sloan[3], Hanlee Ji[2], Philipp Altrock[1], and Noemi Andor[1(✉)]

[1] Integrated Mathematical Oncology, Moffitt Cancer Center, Tampa, FL, USA
noemi.andor@gmail.com
[2] School of Medicine, Stanford University, Stanford, CA, USA
[3] Case Comprehensive Cancer Center, Cleveland Institute for Computational Biology, Case Western Reserve University School of Medicine, Cleveland, OH, USA

Abstract. Gliomas are tumors that evolve from glial cells in the brain or spine. Most gliomas are diagnosed as either lower-grade lesions (grade II) or Glioblastoma (grade IV). Progression of lower-grade gliomas (LGG) to Glioblastoma (GBM) is accompanied by a phenotypic switch to a highly invasive tumor cell phenotype. Converging evidence from different cancer types, including colorectal-, breast-, and lung- cancers, suggests a strong enrichment of high ploidy cells among metastatic lesions as compared to the primary tumor [1,2]. Even in normal development: trophoblast giant cells - the first cell type to terminally differentiate during embryogenesis - are responsible for invading the placenta and strikingly these cells can have up to 1000 copies of the genome [5]. All this points to the existence of a ubiquitous mechanism that links high DNA content to an invasive phenotype. We formulate a mechanistic Grow-or-go model that postulates higher energy demands of high-ploidy cells as a driver of their invasive behavior. We will test whether this mechanism may contribute to the quick recurrence of GBMs after surgery [7] and whether it can explain striking differences in the prognostic power of integrin signaling and cell cycle progression between males and females [13].

Keywords: Ploidy · Glioblastoma · Mathematical modeling

Introduction

Glioblastoma multiforme (GBM) is a devastating brain cancer that invades the brain parenchyma and perivascular space. The moniker "multiforme" is based on first histopathologic descriptions of the diverse morphologic features of the tumor. GBM is a highly hecterogeneous tumor, in which hypercellular regions coexist with areas of interspersed tumor cells spreading beyond the tumor's invasive edge. Intra-tumour genetic heterogeneity emerging from the variable

Research reported in this publication was supported by the National Cancer Institute of the National Institutes of Health under Award Number R00CA215256.

G. Bebis et al. (Eds.): ISMCO 2019, LNCS 11826, pp. 29–34, 2019.
https://doi.org/10.1007/978-3-030-35210-3_2

clonal selection processes within these spatially distinct niches allow glioma cells to evade therapy.

Mathematical Go-or-Grow models have been formulated to describe this dynamic process and to better understand what facilitates the invasive phenotype of GBM [7,10]. These models take into account the ability of proliferative cells to become invasive under certain conditions, such as hypoxia. Here we describe a new Grow-or-go model that postulates higher energy demands of high-ploidy cells as a driver of their invasive behavior. A key similarity to previously described Grow-or-go models is that the switch from highly proliferative to invasive phenotype is modeled as a function of the environment. What is novel here is that our model takes into account a potential contribution of the genotype, namely of large-scale CNVs, to the propensity of that switch.

Evidence supports a stronger role for CNVs rather than SNVs in developing and maintaining population diversity [3,4,8]. Large scale CNVs are so powerful, affect so many genes, even the size of the cell. Gene duplication in particular is arguably the most important evolutionary force on the organism level [11]. Without extensive CNVs, diversity of a population is likely to be limited for rapid adaptation when exposed to stressful environments [3]. CNVs have a locus-specific mutation frequency that is 2 to 4 orders of magnitude greater than that of point mutations [6], and it's precisely this magnitude which makes CNVs so phenotypically effective. However, in contrast to point mutations, CNVs do not offer the convenience of the infinite site assumption, making them more challenging to model. We propose that large-scale CNVs, such as those introduced by chromosome segregation errors, offer many routes for a cell to switch back and forth between migration and proliferation, conferring an underappreciated phenotypic plasticity to the population.

1   Results

1.1   PDE Model

To understand how the distribution of energy (E) influences cell ploidy (c) we model energy and cell density (u) with continuity equations:

$$\frac{\partial \rho}{\partial t} + \nabla \cdot J = q. \tag{1}$$

Here ρ is the quantity of interest (E or u), J is the flux and q is known as a source/sink term. For the energy component, we have two terms for the flux:

$$J_E = J_{\text{E, diff}} + J_{\text{E, adv}} = -\Gamma_E \nabla E + \kappa_E \eta E, \tag{2}$$

where Γ_E is the diffusion coefficient (i.e. diffusion acts along the gradient of energy, in opposing direction), κ_E a measure of the ability for energy to move due to the flow velocity η. The cell density u is a function of ploidy $u(c)$, and we add a ploidy-dependent chemotaxis term $\chi u \nabla g(E, c)$,

$$J_u = J_{\text{u, diff}} + J_{\text{u, adv}} + J_{\text{chemotaxis}} = -\Gamma_u \nabla u + \kappa_u \eta u + \chi u \nabla g(E, c). \tag{3}$$

where χ is the chemotaxis coefficient. Inserting the fluxes into Eq. (1) yields our coupled advection-diffusion equation for the energy and cell concentrations:

$$\frac{\partial E}{\partial t} = \Gamma_E \nabla^2 E - \kappa_E \nabla \cdot (\boldsymbol{\eta} E) + f, \tag{4}$$

$$\frac{\partial u}{\partial t} = \Gamma_u \nabla^2 u - \kappa_u \nabla \cdot (\boldsymbol{\eta} u) - \chi \nabla \cdot [u \nabla g(E)] + h. \tag{5}$$

These equations are valid in the interior of our specified domain (e.g. a circular dish of radius R for cells grown in culture). Here f is related to the consumption of E due to u, and h is related to the birth/death process governing u. Along the boundary, we assume no-flux boundary conditions, which reduces to $\frac{\partial E}{\partial n} = \frac{\partial u}{\partial n} = 0$, where n is the out-pointing normal direction.

1.2   Agent-Based Model

Let x_0 be a cell with proliferation rate α_{x0}, deathrate γ_{x0}, motility rate η_{x0} and ploidy c_{x0}. $S_{x0}^n := \{x_0, x_1, \ldots, x_{k_n}\}$, are living cells in the neighborhood of x_0 up to reachability level n, where $k_n \leq K_n := 8 \cdot \frac{n \cdot (n+1)}{2}$. The probability of observing a cell division, given we consider x_0, is proportional to the energy, E_{x0}, available to the cell:

$$P(\alpha|x0) \sim o \cdot \alpha_{x0} \cdot \text{sgn}\,(E_{x0}) \tag{6}$$

where $o := \begin{cases} 1, & \text{if } k_1 < K_1 \\ 0, & \text{otherwise} \end{cases}$ and $E_{x0} = E \cdot K_n - \sum_{x \in S_{x0}^n} c_x - \frac{1}{2} c_{x0} \cdot K_n$

$E \cdot K_n$ is the maximum energy available for any given cell. The higher the neighborhood cell density and the higher the ploidy, the more likely it is that the cell's energy-demands for division are not met by the environment. This comparison between available and required energy is a surrogate for the dual role of integrin signaling: integrin-mediated signals allow cells to progress from G1 to S phase of the cell cycle. At the same time integrins mediate cell migration. The probability of observing movement while tracing x_0, is proportional to:

$$P(\eta|x_0) \sim o \cdot \alpha_{x0} \cdot \text{sgn}\,(-E_{x0}) \tag{7}$$

i.e. when insufficient energy prevents a cell from dividing, it is more likely to migrate. Insufficient energy for proliferation implicitly increases motility.

The model was implemented as a cellular automaton whereby energy is initially uniformly distributed across a 150-by-150 grid. Any given cell can access energy from its neighborhood, thereby decreasing the available energy of that neighborhood accordingly. Every time a cell dies, it releases the same amount of energy in its neighborhood as it requires for cell division. At each timestep actively migrating and dividing cells are chosen at random according to their probabilities. Candidate new locations are all free locations in the Moore neighborhood of an active cell.

Final location of each migrating cell is chosen among all candidate locations, such that the new location is towards a less dense neighborhood. Final location of each dividing cell is randomly chosen among all candidates. A dividing cell gains/loses a copy of its chromosomes to its daughter cell at a rate v. The effect of the mutation on the fitness of both daughter cells is modelled as a change in the cells' proliferation- and death-rates:

$$\alpha_{x0}^* := \alpha_{x0} \cdot \mathrm{N}\left(1, \sigma^2\right) \tag{8}$$

$$\gamma_{x0}^* := \gamma_{x0} \cdot \mathrm{N}\left(\frac{2}{c_{x0}}, \kappa^2\right), \tag{9}$$

i.e. the copy-losing and -gaining daughter cell will have decreased and increased robustness to future mutations respectively.

We ran 2,500 simulations at variable energies, each time allowing cells to evolve over 100 time steps (Table 1). In low-energy environments high-ploidy clones were enriched at the leading edge of the tumor. This was not the case in high-energy environments (Fig. 1). Jupyter notebook to re-run the simulations and subsequent analyses is available at https://github.com/MathOnco/GoOrGrow_PloidyEnergy.

Table 1. Initial conditions and simulation parameters.

Description	Name	Value or Range
Cell division	α	0.32
Missegregation rate	v	0.02
Cell death	γ	0.25
Cell motility	η	0.2
Ploidy	c	2.5
Energy	E	100–2,200

2   Discussion

CNVs are a phenotypically effective form of genomic instability, leading to changes in the expression of a lot of genes simultaneously, even affecting the size of the cell. Yet most studies have focused on point mutations for the convenience that comes with the infinite-site assumption. Our model proposes chromosome missegregations and the resulting CNVs as an efficient route for cells to switch between migration and proliferation. Aneuploidy rates have been shown to vary between chromosomes after drug-exposure [12], motivating future modelling of chromosome-specific segregation rates.

Future validation experiments will leverage spatially annotated H&E slides from different regions of a GBM from the Ivy GBM atlas [9]. This will include comparing the size and staining intensity of nuclei between regions annotated as "cellular tumor" vs. "leading tumor edge". A higher representation of larger and

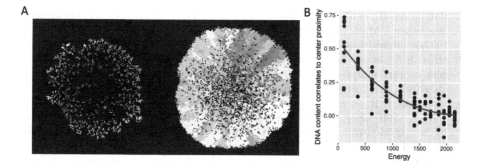

Fig. 1. Agent-based model. (**A**) Two representative simulations are shown. Every dot is a cell growing in either one of two static environments of uniformly distributed low energy (left) or high energy (right). Cells are colored by ploidy - from dark-red (low ploidy) to yellow (high ploidy). (**B**) In low-energy environments we see high-ploidy clones enriched at the leading edge of the tumor. This is not the case in high-energy environments: when energy falls below a certain threshold, a cell's ploidy starts to correlate with its distance from the tumor core, up to a correlation coefficient of 0.5. (Color figure online)

darker nuclei in the latter would be in line with the prediction of our model and therefor support its assumptions. As a second line of validation we will expose GBM cell lines to different flow speeds and use the corresponding response of the fixation probability of high-ploidy clones as an indicator for model selection.

References

1. Angelova, M., et al.: Evolution of metastases in space and time under immune selection. Cell **175**(3), 751–765.e16 (2018). https://doi.org/10.1016/j.cell.2018.09.018. ISSN: 0092-8674, sciencedirect.com/science/article/pii/S0092867418312303. Accessed 15 Mar 2019
2. Brastianos, P.K., et al.: Genomic characterization of brain metastases reveals branched evolution and potential therapeutic targets. Cancer Discovery **5**(11), 1164–1177 (2015). https://doi.org/10.1158/2159-8290.CD-15-0369. ISSN: 2159-8290
3. Chen, G., et al.: Hsp90 stress potentiates rapid cellular adaptation through induction of aneuploidy. Nature **482**(7384), 246–250 (2012). https://doi.org/10.1038/nature10795. ISSN: 1476-4687
4. Chen, G., et al.: Targeting the adaptability of heterogeneous aneuploids. Cell **160**(4), 771–784 (2015). https://doi.org/10.1016/j.cell.2015.01.026. ISSN: 1097-4172
5. Hannibal, R.L., et al.: Copy number variation is a fundamental aspect of the placental genome. PLoS Genet. **10**(5), e1004290 (2014). https://doi.org/10.1371/journal.pgen.1004290. ISSN: 1553-7404
6. Hastings, P.J., et al.: Mechanisms of change in gene copy number. Nat. Rev. Genet. **10**(8), 551–564 (2009). https://doi.org/10.1038/nrg2593. ISSN: 1471-0056, ncbi.nlm.nih.gov/pmc/articles/PMC2864001/. Accessed 15 Aug 2015

7. Hatzikirou, H., et al.: 'Go or grow': the key to the emergence of invasion in tumour progression? Math. Med. Biol. J. IMA **29**(1), 49–65 (2012). https://doi.org/10.1093/imammb/dqq011. ISSN:1477-8602

8. Mroz, E.A., et al.: Intra-tumor genetic heterogeneity and mortality in head and neck cancer: analysis of data from the Cancer Genome Atlas. PLoS Med. **12**(2), e1001786 (2015). https://doi.org/10.1371/journal.pmed.1001786. ISSN: 1549-1676

9. Puchalski, R.B., et al.: An anatomic transcriptional atlas of human glioblastoma. Science **360**(6389), 660–663 (2018). https://doi.org/10.1126/science.aaf2666. ISSN: 0036-8075, 1095-9203, science.sciencemag.org/content/360/6389/660. Accessed 15 Mar 2019

10. Saut, O., et al.: A multilayer grow-or-go model for GBM: effects of invasive cells and anti-angiogenesis on growth. Bull. Math. Biol. **76**(9), 2306–2333 (2014). https://doi.org/10.1007/s11538-014-0007-y. ISSN: 1522-9602

11. Soukup, S.W., Ohno, S.: Evolution by Gene Duplication, p. 160. Springer, New York (1970). Teratology **9**(2), 250–251 (1974). ISSN: 1096-9926, https://doi.org/10.1002/tera.1420090224, http://onlinelibrary.wiley.com/doi/10.1002/tera.1420090224/abstract. Accessed 31 Aug 2014

12. Worrall, J.T., et al.: Non-random mis-segregation of human chromosomes. Cell Rep. **23**(11), 3366–3380 (2018). https://doi.org/10.1016/j.celrep.2018.05.047. ISSN: 2211-1247

13. Yang, W., et al.: Sex differences in GBM revealed by analysis of patient imaging, transcriptome, and survival data. Sci. Trans. Med. **11**(473), eaao5253 (2019). https://doi.org/10.1126/scitranslmed.aao5253. ISSN: 1946-6234, 1946-6242, stm.sciencemag.org/content/11/473/eaao5253. Accessed 15 Mar 2019

Imaging and Scientific Visualization for Cancer Research

cmIF: A Python Library for Scalable Multiplex Imaging Pipelines

Jennifer Eng[✉], Elmar Bucher, Elliot Gray, Lydia Grace Campbell,
Guillaume Thibault, Laura Heiser, Summer Gibbs, Joe W. Gray,
Koei Chin, and Young Hwan Chang

Biomedical Engineering and Center for Spatial Systems Biomedicine,
Oregon Health and Science University, Portland, OR, USA
engje@ohsu.edu

Abstract. Histological staining and analysis of tissue sections is integral to diagnosis and treatment of many diseases, including cancer. Multiplex imaging technologies (e.g., cyclic immunostaining) have dramatically increased capabilities for assessing prognostic biomarkers *in situ*, enabling new insights into complex diseases. However, high-resolution, multiplex image data can be terabytes (TB) in size, and traditional pipelines for image analysis are not suited for these rich datasets. While much software development effort goes towards improving image processing tools such as stitching, registration, and segmentation; integration of these tools into a pipeline is often manual, which is highly laborious, error-prone and lacks reproducibility and scalability. Therefore, we developed a Python3 library, cmIF, a free and open-source tool to handle our high-throughput multiplex image processing pipeline. cmIF enables analysis of full-slide pathology tissue sections and tissue microarrays (TMAs), facilitating processing from raw image files through registration, segmentation, feature extraction, manual thresholding, and spatial pattern analysis. Our cmIF library includes functionality for image handling, quality control, metadata extraction, and subtraction of background images (i.e., autofluorescence subtraction). Additionally, it includes a Jupyter notebook for efficient generation and visualization of manual thresholds. Compared to a manual pipeline, use of cmIF reduces errors and improves processing time of datasets from weeks to hours, while documenting processing steps for reproducibility. All code is available on https://gitlab.com/engje/cmif. While our library is specific to our pipeline elements, it is a blueprint for types of functions needed for high throughput analysis. In the future, we will continue developing this open-source tool, and with input from the wider community, adapt it to a range of multiplex image pipelines.

Keywords: Multiplex imaging · Image processing · High-throughput analytics

---

J.W. Gray, K. Chin, Y.H. Chang—Contributed equally.

G. Bebis et al. (Eds.): ISMCO 2019, LNCS 11826, pp. 37–43, 2019.
https://doi.org/10.1007/978-3-030-35210-3_3

1 Introduction

Staining, imaging, and analysis of tissue sections reveals cell phenotypes in native *in vivo* context, elucidating normal and diseased tissue biology. Multiplex imaging enables identification of numerous cell types and their spatial relationships, with the objective of informing treatment in diseases whose progression is impacted by numerous cell-cell interactions, such as cancer [1]. Recently developed cyclic immunostaining procedures detect dozens of markers in individual tissue sections through repeated staining, imaging, and signal removal [2–7]. Such techniques produce rich, high-resolution image datasets; however, the data pose unique challenges for image processing and analysis. For example, using automated scanning systems, groups, including ours, routinely acquire datasets with more than 60 high-resolution scans, totaling terabytes (TB) in size for one experiment. Existing tools for image analysis are not suited for the size and number of images acquired.

Challenges include lengthy image-opening and viewing operations and repetitive but error-prone tasks. A typical pipeline includes stitching, registration, background subtraction, segmentation, feature extraction, cell classification, and visualization. Processing pipelines must integrate modules developed in different languages (Java, Matlab, Python, custom software, etc.) and this integration is often manual. Additional challenges include accurate capture of metadata, including stain information and microscope settings. Finally, quality control (QC) is an important consideration as errors occur during the manual stain or scan setup steps, while image quality deviations lead to failure of registration, segmentation or background subtraction.

Herein we present an open-source Python library, cmIF, as a solution to the issues outlined above. Even as new methods such as deep learning are leveraged for improved image registration or segmentation, less visible issues such as automated pipeline implementation, QC and data visualization plague many users. Our open-source tools for automated processing, QC, and documentation enable high-throughput, reproducible analysis of multiplex imaging data.

2 Implementation

Our cmIF library for high-throughput, reproducible processing and analysis of multiplex imaging data is implemented in Python 3. There are three major components. The first, mpimage, is a collection of functions that act directly on the image files. Scikit-image is used to load and write 16-bit image files (https://scikit-image.org). Images can be arrayed side by side, overlaid or subtracted. Image attributes including biomarker name, channel, and exposure time are automatically extracted. For example, biomarker and channel are obtained with functions to parse different file-naming schemes (parse_org, parse_img). Exposure time and light intensity are extracted from microscope files using the Python bioformats library to read proprietary image formats (get_exposure) (https://github.com/CellProfiler/python-bioformats). One of the main innovations of this module is flexible implementation of an autofluorescence subtraction step (subtract_images). While this is conceptually simple, i.e. simple subtraction of a blank

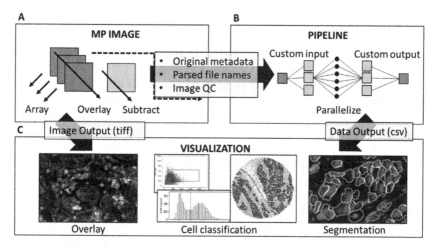

Fig. 1. cmIF Functionality **A.** *Multiplex Image Handling*. Functions array, overlay and subtract grayscale images and extract image metadata including biomarker name, channel, and exposure time, for QC and downstream analysis. **B.** *Pipeline*. cmIF integrates separate image processing steps into an automated pipeline. From image/metadata inputs, cmIF automatically generates custom inputs for each module and starts batch/parallel processing. **C.** *Visualization*. A Jupyter notebook enables the user to efficiently set manual intensity thresholds and visualize the results.

channel image from a stained image [7], this is often missing from multiplex workflows [3, 5, 9].

The second component is a collection of functions to automate image processing pipelines. The preprocess module performs QC, prepares registration and segmentation inputs and starts parallel jobs on a server. Following registration, segmentation and feature extraction, the getdata module formats data into.csv's and the process module filters data for cells with complete retention across rounds, normalizes each channel by exposure time, performs autofluorescence subtraction, and prepares files for thresholding. We use the numpy and pandas libraries for these operations (https://numpy.org, https://pandas.pydata.org).

The last component is a Jupyter notebook for manual thresholding and visualization of results. We elected to develop our own thresholding solution because available software had size and speed limitations (e.g. 500 × 500 pixels) [10]. While manual gating is the gold standard for cell classification, many published analyses using multiplex imaging data avoid gating in favor of clustering [5, 11]. We found current clustering approaches to be highly sensitive to imaging artifacts (edge effect, areas of photo bleaching) as well as negative staining. Therefore, we developed a Jupyter notebook to efficiently set manual thresholds on thousands of images. Matplotlib is used for plotting (https://matplotlib.org).

3   Results

To illustrate the functionality of our library, we processed a medium-sized dataset comprised of five core-needle biopsy samples, acquiring full scans of twelve 5-channel rounds of staining, and two 5-channel background rounds. In total, we generated data for 48 biomarkers and 8 background images in each tissue sample.

3.1   Quality Control

In our example dataset, we used cmIF to perform quality control. We identified the following QC issues in a total of 685 grayscale images from 137 separate 5-channel scans: swapped images (3), missing images (4), poor cropping of tissue borders (2), out-of-focus scans (3), poorly-exposed scans (1), mis-registration (2) and non-standard biomarker names (4). While the rates of manually (17/19) and algorithmically (2/19) introduced errors can differ between operators and datasets, it is important to identify potential errors while documenting their occurrence and correction. cmIF accomplishes this in minutes, while a manual check of the images would take significantly longer and might fail to detect problems.

3.2   Metadata Extraction and Autofluorescence Subtraction

Metadata tracking is an important part of pipeline functionality. Although imaging systems store imaging parameters in proprietary file formats, this information is lost upon export of images to standard formats, i.e. tiff, for processing. Using cmIF, we read metadata directly from proprietary files and extracted the desired information. In our dataset, we read 355 exposure times from Zeiss.czi's, and found one differing from the manually recorded time. Exposure times impact scaling for background subtraction as well as normalization for quantitative comparison across datasets.

Following acquisition of the correct exposure times, we scaled each image by exposure time and performed autofluorescence subtraction. Visual inspection of the images allowed us to determine the optimal channels to use as background, which could be either quenched images (N = 3), failed stains (N = 3) or none (N = 1). Importantly, cmIF allows complete control over specification of which channel should be designated as background for each individual channel/marker, and documents this choice.

3.3   Pipeline Operations

In our case and others, it is necessary to use custom algorithms for stitching, registration and segmentation. Our workflow uses in-house Matlab code for registration and an in-house Java-based software for single-cell segmentation and feature extraction. cmIF is specific to these elements but can be modified to integrate other custom algorithms.

Our pipeline starts with registration of images generated by repeated rounds of multi-channel imaging. Previously it was necessary to manually edit and run a separate Matlab script to register images for each tissue. cmIF automatically edits these scripts and starts the registration jobs in parallel. It also documents all parameters, such as coordinates for tiling large (>4 GB) images and the reference round for registration.

Single-cell segmentation and feature extraction, similar to registration, previously required the manual input of each biomarker name and subcellular localization, staining round, microscope channel and exposure time, and intensity threshold. These inputs had to be prepared separately for each tissue, which became unmanageable as we scaled to dozens and hundreds of tissues. cmIF's preprocess module automatically extracts metadata, including biomarker name and exposure time, and generates the necessary documents for each segmentation job. Another function allows rapid screening and documentation of intensity thresholds (check_seg_markers). Automation of these tasks dramatically improves efficiency and reduces errors. Following segmentation and feature extraction, we convert custom outputs into our standardized output, which is returned to cmIF for downstream processing (Fig. 1B).

3.4 Visualization and Cell Classification

An important problem following extraction of single-cell features is cell classification. We created a Jupyter notebook allowing interactive visualization of manual gating results. Due to the inherent limitations in displaying full resolution image data in real time, we opted to visualize cell identity and location in tissue as a scatterplot (Fig. 1C). This allows thresholds to be quickly updated and visualized across dozens of markers and tissues. Our scatterplot visualizations may be combined with existing image viewers (ImageJ, Zeiss Zen, etc.) for visualization of images. Another feature facilitating fast visualization is a function in mpimage that converts images to 8 bit depth and makes custom overlays of multipage tiffs that can be viewed with standard software (Fig. 1C).

3.5 Scalability

To demonstrate the power of our automated processing pipeline, we analyzed a large dataset comprised of 424 breast cancer tissues arrayed in four separate tissue microarrays (TMAs). The total number of images collected in just two experiments was over 23,000. While manual processing and QC of our previously discussed biopsy dataset is possible, this TMA dataset illustrates the need for automated processing. We performed automated QC on the images and after excluding all tissues with any tissue loss or other imaging issues over 11 rounds of staining, we went on to analyze 163 tissues. Using our Jupyter notebook for efficient manual gating, we set thresholds on 3,586 separate images, and generated single cell positive/negative calls for 1,185,286 total cells. While processing took $\sim$3 weeks of total time for one student, generation of such a large dataset would not be possible without an automated pipeline.

3.6 Code Availability

All source code and documentation, and an example pipeline scripts and dataset are available free and open-source under GPL v3 license at https://gitlab.com/engje/cmif.

4 Discussion

We offer a solution that integrates and automates multiplex image analytics to obtain single-cell classification and localization for deep understanding of normal and diseased tissues. Our lab has roughly 5 years of experience with generation of multiplex images using the cyclic immunofluorescence method [2, 3] and 2 years of experience with image processing, including a transition from a manual to an automated pipeline. Accordingly, we offer tools addressing often ignored, non-trivial problems such as automation, QC, background subtraction, incorporation of custom algorithms, and efficient visualization.

Limitations of our pipeline include the necessity for familiarity with Python and the command line interface. However, this enables programmatic control of highly repetitive processes and facilitates better documentation and reproducibility than current software solutions (https://www.akoyabio.com, http://www.qi-tissue.com). Other limitations include the need for users to customize functions for their own datasets, e.g. to parse filenames or include other registration or segmentation algorithms. However, we include functions in our mpimage module that work on different grayscale image types, as well as a general filename parsing function that has been successfully adopted by collaborators. We expect that groups adopting our software tool will experience improved workflow efficiency and achieve their scientific aims faster and more easily.

Financial Supports. NIH/NCI U54 CA209988, NIH/NCI U2C CA233280, Prospect Creek Foundation, Susan G. Komen Foundation, OHSU Foundation, and Oregon Clinical & Translational Research Institute.

References

1. Hanahan, D.: A: hallmarks of cancer: the next generation. Cell **144**(5), 646–674 (2011)
2. Eng, J., Thibault, G., Luoh, S.-W., Gray, J.W., Chang, Y.H., Chin, K.: Cyclic multiplexed-immunofluorescence (cmIF), a highly multiplexed method for single-cell analysis. In: Thurin, M., Cesano, A., Marincola, Francesco M. (eds.) Biomarkers for Immunotherapy of Cancer. MMB, vol. 2055, pp. 521–562. Springer, New York (2020). https://doi.org/10.1007/978-1-4939-9773-2_24
3. Lin, J.-R.: A simple open-source method for highly multiplexed imaging of single cells in tissues and tumors. ELife, 7 (2018)
4. Tsujikawa, T.: Quantitative multiplex immunohistochemistry reveals myeloid-inflamed tumor-immune complexity associated with poor prognosis. Cell Rep. **19**(1), 203–217 (2017)
5. Goltsev, Y.: Deep profiling of mouse splenic architecture with resource deep profiling of mouse splenic architecture with CODEX multiplexed imaging. Cell **174**(4), 968–981 (2018)
6. McKinley, E.T.: Optimized multiplex immunofluorescence single-cell analysis reveals tuft cell heterogeneity. JCI Insight **2**(11), 93487 (2017)
7. Gut, G.: Multiplexed protein maps link subcellular organization to cellular states. Science **361**, 7042 (2018)
8. Pang, Z.: Dark pixel intensity determination and its applications in normalizing different exposure time and autofluorescence removal. J. Microsc. **246**, 1–10 (2012)

9. Czech, E.: Cytokit: a single-cell analysis toolkit for high dimensional fluorescent microscopy imaging. BMC Bioinformatics **20**(448), 1–13 (2018)
10. Schapiro, D.: histoCAT : analysis of cell phenotypes and interactions in multiplex image cytometry data. Nat. Methods **14**(9), 873 (2017)
11. Keren, L.: A Structure tumor-immune microenvironment in triple negative breast cancer revealed by multiplexed ion beam imaging. Cell **174**(6), 1373–1387.e19 (2018)

Statistical Methods and Data Mining for Cancer Research (SMDM)

Accurate and Flexible Bayesian Mutation Call from Multi-regional Tumor Samples

Takuya Moriyama[1], Seiya Imoto[2], Satoru Miyano[1,2],
and Rui Yamaguchi[1,3,4(✉)]

[1] Human Genome Center, The Institute of Medical Science,
The University of Tokyo, Tokyo, Japan
moriyama@hgc.jp, {imoto,miyano}@ims.u-tokyo.ac.jp
[2] Health Intelligence Center, The Institute of Medical Science,
The University of Tokyo, Tokyo, Japan
[3] Division of Cancer Systems Biology, Aichi Cancer Center Research Institute,
Nagoya, Japan
r.yamaguchi@aichi-cc.jp
[4] Department of Cancer Informatics,
Nagoya University Graduate School of Medicine, Nagoya, Japan

Abstract. We propose a Bayesian method termed MultiMuC for accurate detection of somatic mutations (mutation call) from multi-regional tumor sequence data sets. To improve detection performance, our method is based on the assumption of mutation sharing: if we can predict at least one tumor region has the mutation, then we can be more confident to detect a mutation in more tumor regions by lowering the original threshold of detection. We find two drawbacks in existing methods for leveraging the assumption of mutation sharing. First, existing methods do not consider the probability of the "No-TP (True Positive)" case: we could expect mutation candidates in multiple regions, but actually, no true mutations exist. Second, existing methods cannot leverage scores from other state-of-the-art mutation calling methods for a single-regional tumor. We overcome the first drawback through evaluation of the probability of the No-TP case. Next, we solve the second drawback by the idea of Bayes-factor-based model construction that enables flexible integration of probability-based mutation call scores as building blocks of a Bayesian statistical model. We empirically evaluate that our method steadily improves results from mutation calling methods for a single-regional tumor, e.g., Strelka2 and NeuSomatic, and outperforms existing methods for multi-regional tumors through a real-data-based simulation study. Our implementation of MultiMuC is available at https://github.com/takumorizo/MultiMuC.

1 Introduction

The process of genomic alteration is one of the most important factors for carcinogenesis. Acquired somatic mutations, together with individual germline variations, have a large effect on cancer evolution. By obtaining accurate genomic

© Springer Nature Switzerland AG 2019
G. Bebis et al. (Eds.): ISMCO 2019, LNCS 11826, pp. 47–61, 2019.
https://doi.org/10.1007/978-3-030-35210-3_4

alteration profiles, we can estimate the cause of cancer for individual patients and search for optimal therapies. Thus, mutation calling from sequence data sets has become a fundamental analysis in cancer therapy and research. An enormous number of studies [1–8] have been conducted to improve the performance of single-tumor-based mutation call, i.e., mutation call from a tumor and a matched normal sequence data set, and the performance of mutation call is updated annually by modeling properties of raw sequence data sets in more sophisticated manners. Strelka2 and OHVarfinDer construct Bayesian statistical models to utilize sequence data specific properties. DeepVariant [9] is a convolutional neural network (CNN) based method for detecting germline mutations and able to learn the properties in any sequence data platform. NeuSomatic is also a CNN based method for somatic mutation call, which is motivated by DeepVariant.

Mutation profiles from multi-regional tumor sequencing data sets give helpful information to understand the tumor evolutionary process and the intratumoral heterogeneity. In order to detect subclonal mutations with lower variant allele frequencies, researchers have developed mutation calling methods that are suitable for multi-regional tumor data sets. There are mainly two types of approaches for multi-regional mutation call. The first type of the methods [10–13] consider the property of tumor phylogenetic tree and clonal populations. The second type [14] focused on the sharing assumption of mutation across multiple samples, defined in Sect. 2.1. For these multi-regional mutation calling methods, comprehensive performance evaluations were conducted in recent reports [15].

Although one of the existing methods of multiSNV is based on the sharing assumption of mutation and improved the performance of mutation call, there are still two drawbacks. First, multiSNV does not consider the "No-TP case": even if we could detect mutation candidates in multiple regions, no true mutations exist, unfortunately. We will define No-TP case in Sect. 2.2. Second, detection of a mutation for each tumor region in multiSNV is based on scores from a set of pre-defined generative models and cannot leverage scores from other state-of-the-art mutation calling methods for a single-regional tumor.

Here, we propose a Bayesian method of MultiMuC for multi-regional mutation call. Our method has two defining characteristics. First, our method avoids the No-TP case by leveraging the specificity of detection and the number of detected candidates. We evaluate the probability of the No-TP case and investigate that the probability decreases as the specificity of detection or the number of detected candidates increases. Second, our method can incorporate scores from state-of-the-art mutation calling methods as long as these scores are based on probabilities, i.e., Bayes factors [16] or posterior probabilities. We investigate that Bayes factors provide sufficient information for maximum a posteriori (MAP) estimate even if data generation probabilities for each data set are not available. We demonstrate that our method improves the original detection performance in state-of-the-art mutation calling methods for a single-regional tumor through real-data-based (TCGA 4 mutation calling benchmark datasets) sequence data simulation and outperforms existing multi-regional mutation calling methods.

2   Methods

2.1   The Mutation Sharing Assumption

Here, we explain the mutation sharing assumption that is leveraged to improve the performance of multi-regional mutation call. We assume that there are N sequence data sets $\{D_i\}_{i=1,\cdots,N}$ and latent variables $\{X_i\}_{i=1,\cdots,N}$ ($X_i \in \{0,1\}$) express the existence of a mutation at i-th data set and $C \in \{0,1\}$ represents the existence of the mutation at least one data set and $\{V_i\}_{i=1,\cdots,N}$ ($V_i \in \mathbb{R}$) are the scores from single-tumor-based mutation call. The concept of the mutation sharing assumption for mutation call can be summarized in the following assumption.

Assumption 1 (The mutation sharing assumption)

$\forall v \in \mathbb{R}, \exists w < v \ s.t. \ e(w|C=1) = e(v), \ r(w) > r(v),$

where

$$e(v) := \frac{1}{N}\sum_{i=1}^{N} Pr(X_i = 1|V_i > v) \ \text{(Precision on average)}$$

$$e(v|C=c) := \frac{1}{N}\sum_{i=1}^{N} Pr(X_i = 1|V_i > v, C = c) \ \text{(Precision given C)}$$

$$r(v) := \frac{1}{N}\sum_{i=1}^{N} Pr(V_i > v|X_i = 1) \ \text{(Recall on average)}$$

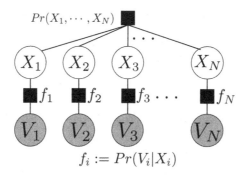

$$f_i := Pr(V_i|X_i)$$

Fig. 1. Graphical representation of the assumed stochastic dependence between $\{X_i\}_{i=1,\cdots,N}$ and $\{V_i\}_{i=1,\cdots,N}$. In this assumed stochastic dependence, we do not set any independence of prior distribution in $Pr(\{X_i\}_{i=1,\cdots,N})$ and only assume that each V_i is dependent on the corresponding X_i.

According to this assumption, if we know (or predict with high confidence) the existence of a mutation in at least one tumor data set ($C = 1$), then we can improve recall from $r(v)$ up to $r(w)$ by lowering the threshold from v down to w with constant precision $e(w|C = 1) = e(v)$. Based on this idea, multiSNV [14] has succeeded in performance improvement. The assumption reflects on the increase of posterior odds as described in Appendix A.1.

2.2   The Probability of the No True Positive (No-TP) Case

We focus on the No-TP cases that cause performance degradation. We define the No-TP case for v detection threshold and M candidate number as the case in which the mutation does not truly exist in any region, even if M candidate mutations are found by the threshold value of v. Under the No-TP case, we cannot obtain any true mutations by lowering the threshold and obtain only false positives instead, thereby negatively affecting the performance of detection. For simplicity, we assume that $V_1 \geq V_2 \cdots \geq V_N$ and we define the probability of the No-TP case as follows.

$$Pr(X_1 = 0, \cdots, X_N = 0 | V_1 > v, \cdots, V_M > v, V_{M+1} \leq v, \cdots, V_N \leq v) \quad (1)$$

To evaluate the probability, we assume the stochastic dependence as shown in the graphical model of Fig. 1. In this setting, we do not set any restriction for stochastic dependence between $X_1, \cdots, X_N$ and only assumes the following conditional independence between $V_1, \cdots, V_N$.

$$Pr(V_1, \cdots, V_N | X_1, \cdots, X_N) = \prod_{i=1}^{N} Pr(V_i | X_i) \quad (2)$$

To sketch our idea, we evaluate the probability of the No-TP case when $M = N$.

$$Pr(X_1 = 0, \cdots, X_N = 0 | V_1 > v, \cdots, V_N > v)$$
$$\propto Pr(X_1 = 0, \cdots, X_N = 0, V_1 > v, \cdots, V_N > v)$$
$$= Pr(X_1 = 0, \cdots, X_N = 0) Pr(V_1 > v, \cdots, V_N > v | X_1 = 0, \cdots, X_N = 0)$$
$$= Pr(X_1 = 0, \cdots, X_N = 0) \prod_{i=1}^{N} Pr(V_i > v | X_i = 0) \quad \because) \ Eq.(2)$$
$$= Pr(X_1 = 0, \cdots, X_N = 0) \prod_{i=1}^{N} (1 - s_i(v)), \quad (3)$$

where $s_i(v) := Pr(V_i \leq v | X_i = 0)$ corresponds to the specificity. Therefore from Eq. (3), we can decrease the probability of the No-TP case by increasing the number of mutation candidates $M (= N)$ or improving the specificity $s_i(v)$. In Appendix A.2, we evaluate the probability in the other case of $M < N$.

2.3 Leveraging Scores from Other Methods for Bayesian Models

We propose an idea to leverage probabilistic scores from other state-of-the-art mutation calling methods for a single-regional tumor to construct a Bayesian hierarchical model for multi-regional tumors.

We can see that data generation probabilities given dependent latent variables can be used as building blocks to construct a Bayesian hierarchical model. For example of Fig. 2, if we can borrow $P(D_i|X_i = \text{Tumor})$ and $P(D_i|X_i = \text{Error})$ as building blocks, then we only need to additionally build the stochastic dependence of latent variables $\{X_i\}_{i=1,\cdots,N}$ to construct the full Bayesian models.

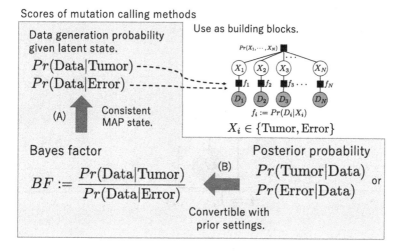

Fig. 2. Summary for usage of probability-based outputs to construct Bayesian models from mutation calling methods for a single-regional tumor.

Although we would like to use the data generation probabilities given dependent latent variables from this idea, e.g., $Pr(\text{Data}|\text{Error})$ and $Pr(\text{Data}|\text{Tumor})$ defined in mutation calling methods for each region of tumor, such probabilities are not available in most cases. On the other hand, alternative values, e.g., Bayes factors or posterior probabilities are available as mutation calling scores from state-of-the-art methods, e.g., Strelka2 and NeuSomatic. In the following sections, we will demonstrate how Bayes factors and posterior probabilities can be used as building blocks to construct a Bayesian model (Fig. 2). First, we will show how to extract equivalent information to the data generation probabilities from Bayes factors by considering maximum a posteriori (MAP) state (Fig. 2A) through introducing a toy example model. Next, we will show how to convert posterior probabilities to Bayes factors (Fig. 2B).

Data Generation Probabilities from Bayes Factors. We show that Bayes factors are sufficient for MAP estimate for a toy example of stochastic models even when full data generation probabilities of $Pr(\text{Data}|\text{Tumor})$ and $Pr(\text{Data}|\text{Error})$ are not given.

For this example, we assume the stochastic model as shown in Fig. 3. $S \in \{0,1\}$ represents the existence of tumor cells and $Y_i \in \{0,1\}$ represents the existence of mutation at the i-th data set D_i. The Bayes factor for the i-th data set is defined as the ratio of the marginal likelihood and P_{call} is defined in a single-tumor-based mutation calling method.

$$BF_i := \frac{P_{call}(D_i|Y_i = 1)}{P_{call}(D_i|Y_i = 0)} \tag{4}$$

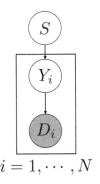

$$i = 1, \cdots, N$$

Fig. 3. A toy example model for multiple tumor samples. $S \in \{0,1\}$ represents the existence of tumor cells and $Y_i \in \{0,1\}$ represents the existence of mutation at i-th data set D_i.

We assume that each data generation probability is a positive value for any observed data point D_i.

$$P_{call}(D_i|Y_i = 1) > 0, P_{call}(D_i|Y_i = 0) > 0 \tag{5}$$

We consider two settings of probability distributions for this toy example model and denote the first setting as $P^{(1)}(\cdot)$ and the second setting as $P^{(2)}(\cdot)$. For both settings, we assume common distributions for S and Y_i.

$$Pr^{(1)}(S) = Pr^{(2)}(S) = Ber(S|f_1)$$
$$Pr^{(1)}(Y_i|S) = Pr^{(2)}(Y_i|S) = Ber(Y_i|f_2)^S \cdot Ber(Y_i|f_3)^{(1-S)}$$

where $Ber(\cdot|f)$ means the probability mass function of Bernoulli distribution with a frequency of f and we set $0 \leq f_1, f_2, f_3 \leq 1$. In the first setting at Eq. (6), we use both the numerator and denominator in each Bayes factor to define the distributions. In the second setting at Eq. (7), we only use Bayes factors and supplement the distributions with a pre-defined positive constant p for all the data index i.

$$Pr^{(1)}(D_i|Y_i = 0) = P_{call}(D_i|Y_i = 0), Pr^{(1)}(D_i|Y_i = 1) = P_{call}(D_i|Y_i = 1) \quad (6)$$

$$Pr^{(2)}(D_i|Y_i = 0) = p, \ Pr^{(2)}(D_i|Y_i = 1) = p \cdot BF_i \ (0 < p) \quad (7)$$

As shown in the following lemma, this difference in setting the probability distribution does not affect the MAP state of the latent variable S and Y_i. Therefore, Bayes factors give sufficient information on data generation probabilities for MAP inference of the latent state for some models of stochastic dependence.

Lemma 1 (Unchanged MAP state).

$$\arg \max_{S,Y} Pr^{(1)}(S, Y|D) = \arg \max_{S,Y} Pr^{(2)}(S, Y|D),$$

where

$$Y := (Y_1, \cdots, Y_N), \ D := (D_1, \cdots, D_N).$$

Proof. It is sufficient if we can show that the following conditions hold true.

· $Pr^{(1)}(S, \{D_i, Y_i\}_i) > 0 \iff Pr^{(2)}(S, \{D_i, Y_i\}_i) > 0$ (Same support region)

· $Pr^{(1)}(S', \{D_i, Y_i'\}_i) > 0, Pr^{(1)}(S, \{D_i, Y_i\}_i) > 0$

$$\Rightarrow \frac{Pr^{(1)}(S', \{D_i, Y_i'\}_i)}{Pr^{(1)}(S, \{D_i, Y_i\}_i)} = \frac{Pr^{(2)}(S', \{D_i, Y_i'\}_i)}{Pr^{(2)}(S, \{D_i, Y_i\}_i)} \quad \text{(Same probability ratio)}$$

The first condition is satisfied from Eq. (5) and $0 < p$.
For the second condition, we can show the condition by substitution. □

Bayes Factors from Posterior Probabilities. Some methods, e.g., Strelka2 and NeuSomatic return the output of posterior event probabilities. We can convert the posterior event probabilities to Bayes factors by setting the prior event ratio, e.g., $Pr(\text{tumor})/Pr(\text{error}) = 1$ used in this paper.

$$BF = \frac{P_{call}(\text{tumor}|D)}{P_{call}(\text{error}|D)} \frac{Pr(\text{error})}{Pr(\text{tumor})} = \frac{P_{call}(\text{tumor}|D)}{1 - P_{call}(\text{tumor}|D)} \frac{Pr(\text{error})}{Pr(\text{tumor})} \quad (8)$$

2.4   Bayesian Statistical Model in MultiMuC

Based on the ideas shown above, we constructed the Bayesian statistical method named as MultiMuC and the graphical summary of MultiMuC is shown in Fig. 4.

Our method is composed of an evidence generation model and a data generation model as shown in the left part and the right part in Fig. 4 respectively. For the evidence generation model, $E_{i,j}$ represents the i-th evidence around the j-th somatic mutation, $Z_{i,j} \in \{0, 1\}$ represents the existence of the j-th somatic

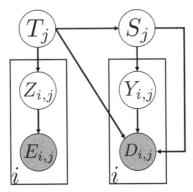

$i = 1, \cdots, N$: Tumor data index.

$j = 1, \cdots, G$: Mutation candidate index.

Fig. 4. Graphical summary of MultiMuC. i represents the location index of tumor sequence data and j represents the index of mutation candidate. The left side of the figure shows the evidence generation model and the right side of the figure shows the data generation model.

mutation at the i-th evidence, and $T_j \in \{0,1\}$ represents the existence of the j-th somatic mutation for at least one evidence. The distributions of these random variables are set as follows.

$$Pr(T_j) = Ber(\cdot|0.5)$$
$$Pr(Z_{i,j}|T_j) = Ber(\cdot|\epsilon)^{1-T_j} \cdot Ber(\cdot|0.5)^{T_j}$$
$$Pr(E_{i,j}|Z_{i,j}) = 1^{1-Z_{i,j}} \cdot H_{i,j}^{Z_{i,j}}$$

$H_{i,j}$ is the Bayes factor and we can detect a mutation with high specificity if $H_{i,j} > 1$. $\epsilon(\approx 0)$ corresponds to the false-positive rate (equal to $1-$specificity) for this Bayes factor of $H_{i,j}$.

For the data generation model, $D_{i,j}$ represents the i-th data set around the j-th mutation candidate, $Y_{i,j} \in \{0,1\}$ represents the existence of the j-th somatic mutation at the i-th data set and $S_j \in \{0,1\}$ represents the existence of the j-th somatic mutation for at least one data set. The distributions of these random variables are set as follows depending on T_j.

$$Pr(S_j|T_j) = Ber(\cdot|0.5)^{1-T_j} Ber(\cdot|p_{con})^{T_j}$$
$$Pr(Y_{i,j}|S_j) = Ber(\cdot|\delta)^{1-S_j} \cdot Ber(\cdot|0.5)^{S_j}$$
$$Pr(D_{i,j}|Y_{i,j}, S_j, T_j) = 1^{1-Y_{i,j}} \left(L_{i,j} \cdot 10^{\theta T_j} \cdot 10^{\rho S_j}\right)^{Y_{i,j}}$$

$L_{i,j}$ is the Bayes factor that is generally used and δ corresponds to its false positive rate. $10^{\theta}(>1)$ lowers the threshold of Bayes factors when the presence of a mutation can be predicted with high specificity ($T_j = 1$). $10^{\rho}(>1)$ also lowers the threshold of Bayes factors when the presence of a mutation can be predicted from the usual result ($S_j = 1$). $p_{con}(\approx 1)$ is the consistency rate from $T_j = 1$ to $S_j = 1$. In this paper, we used the following setting of hyperparameters: $\epsilon = 0.2$, $\delta = 0.02$, $\theta = 0.5$, $\rho = 0.1$ and $p_{con} = 0.999$.

In this method we estimate the MAP state by MCMC [17] for each position j and use $Y'_{i,j}$ for mutation call.

$$\mathbf{Y}'_j, \mathbf{Z}'_j, S'_j, T'_j = \underset{\mathbf{Y}'_j, \mathbf{Z}'_j, S'_j, T'_j}{\arg\max} \; Pr(\mathbf{Y}'_j, \mathbf{Z}'_j, S'_j, T'_j | D_{\cdot,j}, E_{\cdot,j})$$

$$(\mathbf{Y}'_j := (Y'_{1,j}, \cdots, Y'_{N,j}), \; \mathbf{Z}'_j := (Z'_{1,j}, \cdots, Z'_{N,j}))$$

Preparation of Bayes Factors. This method requires Bayes factors with high specificity in addition to the usual Bayes factor results. For preparation of these Bayes factors, we set threshold values and multiplied the original Bayes factor by the inverse number of the threshold as follows.

$$H_{i,j} = BF_{i,j} \cdot 10^{-1.5}, \; L_{i,j} = BF_{i,j} \cdot 10^a, \tag{9}$$

where $BF_{i,j}$ is the original Bayes factor outputs generated by the single-tumor-based method and 10^{-a} corresponds to the general threshold value for the Bayes factors. For mutation calling with high specificity, we set $10^{1.5}$ as the threshold value. For MuTect2, we conducted $a = a - 6.3$ because of the default threshold setting in MuTect2.

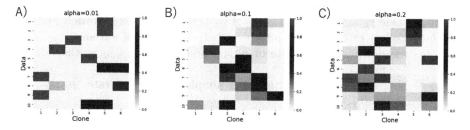

Fig. 5. Examples of simulated clonal mixture rates. (A) illustrates the case of $\alpha = 0.01$ and (B) illustrates the case of $\alpha = 0.1$, and (C) illustrates the case of $\alpha = 0.2$.

3   Results

3.1   Simulation Experiments Based on Real Data Sets

We evaluated MultiMuC performance by simulating multiple tumor sequence datasets. To do this, we used multiple settings for both the tumor phylogenetic tree and the mixture rate of clones, where a clone means a type of tumor cell population. These datasets were prepared in 24 different configurations. Figures 5 and 6 show the examples of the mixture composition rates and tumor phylogenetic trees that were used. The simulation procedures were as follows.

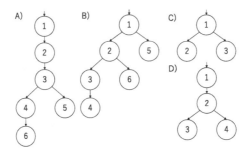

Fig. 6. Simulated trees used for evaluations. Each numbered node corresponds to a clone (a type of tumor cell population) and each edge corresponds to a non-empty set of somatic mutations. For each simulated data set, we sampled a mixture rate of clones and simulated bulk sample data sets.

(1) Collect true somatic mutations and sequence errors from a single pure tumor ($=t_{original}$) and a matched pure normal ($=n_{original}$) data set.
(2) Filter out true mutations with allele frequencies of $<30\%$ or $>70\%$ for allele frequencies to decrease from $\sim50\%$ following the phylogenetic tree.
(3) Generate a random phylogenetic tree $\mathcal{T}$.
(4) Randomly relate each somatic mutation with an edge of the tree $\mathcal{T}$.
(5) For each tumor simulation data set, we generated reads as follows for 10 tumor data sets.
  (5-a) Sample a mixture rate of clones $\boldsymbol{p}_{mix} \sim Dirichlet(\cdot|(\alpha, \cdots, \alpha))$.
  (5-b) For each true somatic mutation s, calculate the total population of clone $p_{tumor} = \sum_{i \in A} p_{mix,i}$, $A := \{i | i\text{-th clone has mutation } s\}$.
  (5-c) Collect reads around the true somatic mutation of s from $t_{original}$ at the down sampling rate of p_{tumor} and from $n_{original}$ at the rate of $1 - p_{tumor}$.
  (5-d) For each error position e, sample an error rate $p_{error} \sim Beta(\cdot|0.1, 0.1)$.
  (5-e) Collect reads around the error position of e from $t_{original}$ at the rate of p_{error} and from $n_{original}$ at the rate of $1 - p_{error}$.

For $t_{original}$ and $n_{original}$, we used real data sets from TCGA 4 mutation calling benchmark datasets (https://gdc.cancer.gov/resources-tcga-users/tcga-mutation-calling-benchmark-4-files).

Performance Comparison. We conducted a performance comparison based on F-measure. We used Strelka2, MuTect2, NeuSomatic and OHVarfinDer as Bayes factor inputs. For the counterpart method, we prepared multiSNV and treeomics. We summarized the F-measures of these methods at $a = 0.0$ in Fig. 7. In this figure, +M indicates that our method was used. Our method steadily contributes to performance improvement for Strelka2, Neu-Somatic and OHVarfinDer and does not cause performance degradation for

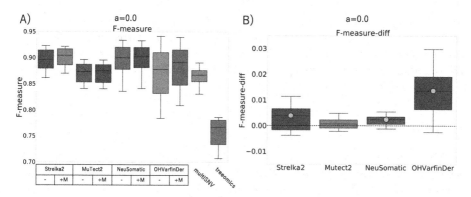

Fig. 7. The summary of F-measures at $a = 0.0$. 10^{-a} is the Bayes factor threshold for mutation call as shown in Eq. (9). +M represents the use of MultiMuC, and an orange-colored circle represents a positive difference of F-measures on average with a P-value less than 0.01 (two-sided paired t-test). (A) represents the summary of F-measure with the threshold at $a = 0.0$. (B) represents the difference of F-measure by applying MultiMuC with the threshold at $a = 0.0$.

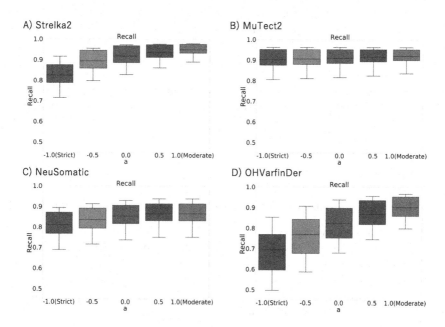

Fig. 8. Summary of recalls in the original mutation calling methods at different default threshold values of 10^{-a}, where $a \in \{-1.0, -0.5, 0.0, 0.5, 1.0\}$. (A) at Strelka2. (B) at MuTect2. (C) at NeuSomatic. (D) at OHVarfinDer.

MuTect2 (Fig. 7B). Furthermore, the combined output of our method and single-tumor-based methods outperformed both multiSNV and treeomics (Fig. 7A). The reason for no statistically significant performance gain in MuTect2 may be due to the increase of recalls at MuTect2 being smaller than that of the other methods used, as shown in Fig. 8.

4   Conclusions

In this paper, we propose a Bayesian method for multi-regional mutation call based on the mutation sharing assumption with two characteristics. First, our method avoids the No-TP case by considering both the specificity of detection and the number of detected candidates to avoid performance degradation. Second, our method can incorporate scores from state-of-the-art mutation calling methods for a single-regional tumor if scores are based on probabilities except for P-values. This improvement for the performance of mutation call will contribute to an improved inference of tumor phylogeny.

For future work, there remains at least one interesting extension of this method to be explored. With this method, we can use the outputs of single-tumor-based methods if posterior event probability, or a Bayes factor, is available. However, our method cannot handle P-value-based outputs of some single-tumor-based methods [1,4,18–20] although P-value is a useful measure for decision making.

Acknowledgments. We used the supercomputers at Human Genome Center, the Institute of Medical Science, the University of Tokyo. This work has been supported by the Grant-in-Aid for JSPS Research Fellow (17J08884) and MEXT/JSPS KAKENHI Grant (15H05912, hp180198, hp170227, 18H03329, hp190158).

A   Appendix

A.1   Increasing Posterior Odds Score of Mutation Call Given $C = 1$

In the main text, we mentioned the assumption that lowering the threshold of scores given $C = 1$ leads to performance improvement at Sect. 2.1. Here, we show that the assumption is based on an increase of posterior odds for mutation call. We assume that each score is represented by posterior odds form $V_i := Pr(X_i = 1|D_i)/Pr(X_i = 0|D_i)$. If we observe $C = 1$ in addition to the observed sequence data set, the true posterior odds can be represented as follows.

$$V_i' := \frac{Pr(X_i = 1|C = 1, D_i)}{Pr(X_i = 0|C = 1, D_i)} = \frac{Pr(X_i = 1, C = 1, D_i)}{Pr(X_i = 0, C = 1, D_i)}$$
$$= \frac{Pr(X_i = 1|C = 1)}{Pr(X_i = 0|C = 1)} \frac{Pr(D_i|X_i = 1, C = 1)}{Pr(D_i|X_i = 0, C = 1)}$$

The true posterior odds is greater than the original posterior odds as shown in the following lemma.

Lemma 2 (Increasing posterior odds).

If $Pr(D_i|X_i, C) = Pr(D_i|X_i)$, $0 < Pr(C = 0) < 1$, and $V_i, V_i' \in \mathbb{R}$, then $V_i' > V_i$.

Proof.

It is sufficient to show that the following condition holds true.

$$\cdot \frac{Pr(X_i = 1|C = 1)}{Pr(X_i = 0|C = 1)} > \frac{Pr(X_i = 1)}{Pr(X_i = 0)}$$

The condition can be proved by evaluating $Pr(X_i = 1)$ and $Pr(X_i = 0)$ as follows.

$$\begin{aligned}
&Pr(X_i = 1) \\
&= Pr(X_i = 1|C = 1)Pr(C = 1) + Pr(X_i = 1|C = 0)Pr(C = 0) \\
&= Pr(X_i = 1|C = 1)Pr(C = 1) \; (\because \; Pr(X_i = 1|C = 0) = 0) \\
&Pr(X_i = 0) \\
&= Pr(X_i = 0|C = 1)Pr(C = 1) + Pr(X_i = 0|C = 0)Pr(C = 0) \\
&= Pr(X_i = 0|C = 1)Pr(C = 1) + Pr(C = 0) \; (\because \; Pr(X_i = 0|C = 0) = 1) \\
&> Pr(X_i = 0|C = 1)Pr(C = 1) \; (\because \; 0 < Pr(C = 0) < 1)
\end{aligned}$$

By using the above evaluations, we can show $\frac{Pr(X_i=1|C=1)}{Pr(X_i=0|C=1)} > \frac{Pr(X_i=1)}{Pr(X_i=0)}$. From this condition and the given hypothesis,

$$V_i' = \frac{Pr(X_i = 1|C = 1)}{Pr(X_i = 0|C = 1)} \frac{Pr(D_i|X_i = 1, C = 1)}{Pr(D_i|X_i = 0, C = 1)} > \frac{Pr(X_i = 1)}{Pr(X_i = 0)} \frac{Pr(D_i|X_i = 1)}{Pr(D_i|X_i = 0)} = V_i$$

$\square$

A.2   The Probability of No-TP Case in General

Here, we evaluate the probability of the No-TP case when $M < N$. For simplicity, we define variables and relational operators between vector and scalar. For Eq. (10), we also define similar relational operators for $\geq, <, \leq, =$ between vector and scalar.

$$\begin{aligned}
&\boldsymbol{V} := (V_1, \cdots, V_M), \, \widetilde{\boldsymbol{V}} := (V_{M+1}, \cdots, V_N), \, \boldsymbol{X} := (X_1, \cdots, X_M), \, \widetilde{\boldsymbol{X}} := (X_{M+1}, \cdots, X_N), \\
&\boldsymbol{u} > v \iff u_i > v \; (\forall i) \\
&\boldsymbol{u} \neq v \iff u_i \neq v \; (\exists i)
\end{aligned} \qquad (10)$$

The probability of the No-TP case can be represented as follows.

$$Pr(\boldsymbol{X} = 0, \widetilde{\boldsymbol{X}} = 0 | \boldsymbol{V} > v, \widetilde{\boldsymbol{V}} \leq v) \qquad (11)$$

For $Pr(\boldsymbol{V} > v, \widetilde{\boldsymbol{V}} \leq v)$, we can obtain a lower bound as follows.

$$
\begin{aligned}
Pr&(\boldsymbol{V} > v, \widetilde{\boldsymbol{V}} \leq v) \\
&= \sum_{\boldsymbol{X},\widetilde{\boldsymbol{X}}} Pr(\boldsymbol{X}, \widetilde{\boldsymbol{X}}) Pr(\boldsymbol{V} > v, \widetilde{\boldsymbol{V}} \leq v | \boldsymbol{X}, \widetilde{\boldsymbol{X}}) \\
&= \sum_{\boldsymbol{X},\widetilde{\boldsymbol{X}}} Pr(\boldsymbol{X}, \widetilde{\boldsymbol{X}}) \prod_{i=1}^{M} Pr(V_i > v | X_i) \prod_{k=M+1}^{N} Pr(V_k \leq v | X_k) \ (\because Eq.\,(2)) \\
&= \sum_{\boldsymbol{X},\widetilde{\boldsymbol{X}}} Pr(\boldsymbol{X}, \widetilde{\boldsymbol{X}}) \prod_{i=1}^{M} (1 - s_i(v))^{1-X_i} R_i(v)^{X_i} \prod_{k=M+1}^{N} s_k(v)^{1-X_k} (1 - R_k(v))^{X_k} \\
&\geq Pr(\boldsymbol{X} = 1, \widetilde{\boldsymbol{X}} = 0) \prod_{i=1}^{M} R_i(v) \prod_{k=M+1}^{N} s_k(v) =: A,
\end{aligned}
\tag{12}
$$

where $R_i(v) := Pr(V_i > v | X_i = 1)$ corresponds to recall.
From Eq. (12), if $A > 0$, we can derive an upper bound for Eq. (11) as follows.

$$
\begin{aligned}
Pr(\boldsymbol{X} = 0, \widetilde{\boldsymbol{X}} = 0 | \boldsymbol{V} > v, \widetilde{\boldsymbol{V}} \leq v) &= \frac{Pr(\boldsymbol{X} = 0, \widetilde{\boldsymbol{X}} = 0, \boldsymbol{V} > v, \widetilde{\boldsymbol{V}} \leq v)}{Pr(\boldsymbol{V} > v, \widetilde{\boldsymbol{V}} \leq v)} \\
&\leq min\left(1, \frac{Pr(\boldsymbol{X} = 0, \widetilde{\boldsymbol{X}} = 0) \prod_{i=1}^{M}(1 - s_i(v)) \prod_{k=M+1}^{N} s_k(v)}{Pr(\boldsymbol{X} = 1, \widetilde{\boldsymbol{X}} = 0) \prod_{i=1}^{M} R_i(v) \prod_{k=M+1}^{N} s_k(v)}\right) \\
&= min\left(1, \frac{Pr(\boldsymbol{X} = 0, \widetilde{\boldsymbol{X}} = 0)}{Pr(\boldsymbol{X} = 1, \widetilde{\boldsymbol{X}} = 0)} \prod_{i=1}^{M} \frac{1 - s_i(v)}{R_i(v)}\right)
\end{aligned}
\tag{13}
$$

From Eq. (13), as the specificity increases, the probability of the No-TP case also decreases when $M < N$.

References

1. Koboldt, D.C., et al.: VarScan 2: somatic mutation and copy number alteration discovery in cancer by exome sequencing. Genome Res. **22**(3), 568–576 (2012)
2. Saunders, C.T., et al.: Strelka: accurate somatic small-variant calling from sequenced tumor-normal sample pairs. Bioinformatics **28**(14), 1811–1817 (2012)
3. Cibulskis, K., et al.: Sensitive detection of somatic point mutations in impure and heterogeneous cancer samples. Nat. Biotechnol. **31**(3), 213–219 (2013)
4. Shiraishi, Y., et al.: An empirical Bayesian framework for somatic mutation detection from cancer genome sequencing data. Nucleic Acid Res. **41**(7), e89 (2013)
5. Usuyama, N., et al.: HapMuC: somatic mutation calling using heterozygous germ line variants near candidate mutations. Bioinformatics **30**(23), 3302–3309 (2014)
6. Kim, S., et al.: Strelka2: fast and accurate calling of germline and somatic variants. Nat. Methods **15**(8), 591–594 (2018)
7. Moriyama, T., et al.: A Bayesian model integration for mutation calling through data partitioning. Bioinformatics, btz233 (2019). https://academic.oup.com/bioinformatics/advance-article/doi/10.1093/bioinformatics/btz233/5423180

8. Sahraeian, S.M.E., et al.: Deep convolutional neural networks for accurate somatic mutation detection. Nat. Commun. **10**(1), 1041 (2019)
9. Poplin, R., et al.: A universal SNP and small-indel variant caller using deep neural networks. Nat. Biotechnol. **36**(10), 983–987 (2018)
10. Reiter, J.G., et al.: Reconstructing metastatic seeding patterns of human cancers. Nature Commun. **8**, 14114 (2017)
11. Dorri, F., et al.: Somatic mutation detection and classification through probabilistic integration of clonal population information. Commun. Biol. **2**(1), 44 (2019)
12. van Rens, K.E., et al.: SNV-PPILP: refined SNV calling for tumor data using perfect phylogenies and ILP. Bioinformatics **31**(7), 1133–1135 (2015)
13. Salari, R., et al.: Inference of tumor phylogenies with improved somatic mutation discovery. J. Comput. Biol. **20**(11), 933–944 (2013)
14. Josephidou, M., et al.: multiSNV: a probabilistic approach for improving detection of somatic point mutations from multiple related tumour samples. Nuclic Acids Res. **43**(9), e61 (2015)
15. Detering, H., et al.: Accuracy of somatic variant detection in multiregional tumor sequencing data. bioRxiv 655605 (2019)
16. Kass, R.E., et al.: Bayes factors. J. Am. Stat. Assoc. **90**(430), 773–795 (1995)
17. Neal, R.M.: Probabilistic inference using Markov Chain Monte Carlo methods. Technical report, Department of Computer Science, University of Toronto (1993)
18. Koboldt, D.C., et al.: VarScan: variant detection in massively parallel sequencing of individual and pooled samples. Bioinformatics **25**(17), 2283–2285 (2009)
19. Wilm, A., et al.: LoFreq: a sequence-quality aware, ultra-sensitive variant caller for uncovering cell-population heterogeneity from high-throughput sequencing datasets. Nuclic Acids Res. **40**(22), 11189–11201 (2012)
20. Narzisi, G., et al.: Genome-wide somatic variant calling using localized colored de Bruijn graphs. Commun. Biol. **1**(1), 20 (2018)

Flexible Data Trimming for Different Machine Learning Methods in Omics-Based Personalized Oncology

Victor Tkachev[1], Anton Buzdin[1,2] (iD), and Nicolas Borisov[1,2(✉)] (iD)

[1] Department of Bioinformatics and Molecular Networks, OmicsWay
Corporation, Walnut, CA, USA
borisov@oncobox.com
[2] I.M. Sechenov First Moscow State Medical University (Sechenov University),
Moscow 119991, Russia

Abstract. Machine learning (ML) methods are still rarely used for gene expression/mutation-based prediction of individual tumor responses on anticancer chemotherapy due to relatively rare clinical case histories supplemented with high-throughput molecular data. This leads to high vulnerability of most ML methods are to overtraining. Recently, we proposed a novel hybrid global-local approach to ML termed FLOating Window Projective Separator (FloWPS) that avoids extrapolation in the feature space and may improve robustness of classifiers even for datasets with limited number of preceding cases. FloWPS has been validated for the support vector machines (SVM) method, where if significantly improved the quality of classifiers. The core property of FloWPS is data trimming, i.e. sample-specific removal of features. The irrelevant features in a sample that don't have significant number of neighboring hits in the training dataset are removed from further analyses. In addition, for each point of a validation dataset, only the proximal points of the training dataset are taken into account. Thus, for every point of a validation dataset, the training dataset is adjusted to form a floating window. Here, we applied this approach to seven popular ML methods, including SVM, k nearest neighbors (kNN), random forest (RF), Tikhonov (ridge) regression (RR), binomial naïve Bayes (BNB), adaptive boosting (ADA) and multi-layer perceptron (MLP). We performed computational experiments for 21 high throughput clinically annotated gene expression datasets totally including 1778 cancer patients who either responded or not on chemotherapy treatments. The biggest dataset had samples for 235, whereas the smallest for 41 individual cases. For global ML methods, such as SVM, RF, BNB, ADA and MLP, FloWPS essentially improved the classifier quality. Namely, the area under the receiver-operator curve (ROC AUC) for the responder vs non-responder classifier, increased from typical range 0.65–0.85 to 0.80–0.95, respectively. On the other hand, FloWPS was shown useless for purely local ML techniques such as kNN method or RR. However, both these local methods exhibited low sensitivity or specificity in cases when false positive or false negative errors, respectively, should be avoided. According to sensitivity-specificity criterion, for all the datasets tested, the best performance in combination with FloWPS data trimming was shown for the binomial naïve Bayesian method, which can be valuable for further development of predictors in personalized oncology.

© Springer Nature Switzerland AG 2019
G. Bebis et al. (Eds.): ISMCO 2019, LNCS 11826, pp. 62–71, 2019.
https://doi.org/10.1007/978-3-030-35210-3_5

Keywords: Bioinformatics · Personalized medicine · Oncology · Chemotherapy · Omics big data · Machine learning

1 Background

Personalized medicine (PM) approach in clinical oncology provides important advantages including improved patient survival and lower side toxicities of drugs [1, 2].

Using transcriptomic data, bioinformatic models can be built for patient-oriented ranking of cancer drugs [3]. These models can be hypothesis-driven, based on the assumptions how drug activity should be connected with molecular features within the tumor [4–6]. Alternatively, non-hypothesis-driven approaches including machine learning (ML) don't need any theory providing a link between the molecular profiles and drug efficiencies but strongly require sufficient training and validation datasets.

Although some ML methods have recently been successfully applied for distinguishing between cancer patients with positive and negative response to certain treatment methods [7–10], even these modern sophisticated techniques were not so successful (AUC < 0.66) in prediction of clinical outcome for large datasets, e.g. for multiple myeloma treatment with bortezomib [11].

From the other side, recently we have suggested another approach to ML [3, 12–15], which is called flexible data trimming. Our data trimming approach avoids it by using the rectangular projections along all irrelevant expression features that cause extrapolation during the ML-based predictions for every validation point. Moreover, for each point of a validation dataset, it takes into account only the proximal points of the training dataset. Thus, for every point of a validation dataset, the training dataset is adjusted to form a floating window, and that is why the ML scheme with flexible data trimming was alternatively called FLOating Window Projective Separator, FloWPS [15].

In the pilot trial [15], FloWPS provided surprisingly high performance for the classifiers: AUC > 0.7 for leave-one-out scheme for all datasets, including the multiple myeloma dataset with 169 patients, where responders and non-responders to bortezomib treatment are generally seem poorly separable using the omics data [11]. In the current work, we will consider its application for other ML methods.

2 Results

2.1 Principles of Flexible Data Trimming

Our data trimming method was applied to classify the clinical response of cancer patients to certain chemotherapy treatment using a training dataset with the gene expression/mutation profiles linked to known response for the same treatment of same disease. Since the number of patients with annotated case histories (when treatment method and its clinical success is known, together with the high-throughput gene expression/mutation profile) is limited, we have tailored the whole data trimming scheme to match the leave-one-out (LOO) methodology.

This LOO approach in our *flexible data trimming* is employed three times [3, 15]:

- first, it helped us to specify the *core marker gene sets*, which form the feature space $\mathbf{F} = (f_1, \ldots, f_S)$ for subsequent application of data trimming;
- second, it was applied for every ML prediction act for the wide range of data trimming parameters, m and k;
- third, it was used for the final prediction of the treatment response for every patient and optimized (for all remaining patients) values of parameters m and k.

Imagine that we have to classify the clinical response for a certain patient i (called *patient of interest*) from a given dataset using the most relevant information, i.e. avoiding extrapolation in the feature space and neglecting the preceding cases, which are too distant from the patient of interest. Let the whole dataset contain N patients, so that the remaining $N–1$ patients form the *preceding dataset* D_i, for the patient of interest. For ML *without data trimming*, in the feature space $\mathbf{F} = (f_1, \ldots, f_S)$ all $N–1$ remaining patients are used to build the classifier.

Let us now trim this *preceding dataset*. First, perform the LOO procedure for prediction of treatment response within the preceding dataset, which has $N–1$ patients. To classify every patient j from the preceding dataset D_i, $N–2$ remaining patients may be used. To avoid extrapolation in the feature space, let us select the subset of $\mathbf{F}_{ij}$ of *relevant features* [15]. A feature f_s is considered relevant for the patient j if on its axis there are at least m projections from $N–2$ training samples, which are larger than $f_s(i,j)$, and, at the same time, at least m, which are smaller than $f_s(i,j)$, when m is a non-negative integer parameter. Note that the relevant subset $\mathbf{F}_{ij}(m)$ is individual for every patient i and j [15].

Next, in the space of relevant features $\mathbf{F}_{ij}(m)$ we keep for training only k samples closest to sample j, from given ($N–2$) patients, except i and j; k is also an integer parameter. As a measure for proximity, the Euclidean distance was used, both previously [15] and now. Thus, the k parameter specifies the number of nearest neighbors in the subspace of selected features. Although this approach seems quite similar to relatively simple k nearest neighbors (kNN) ML method [16], unlike the kNN, where k is relatively small and does not usually exceed 20, for our data trimming method we found reasonable making the k value higher [15].

After selection of relevant features and nearest neighbors for the patient j, the ML model is trained using nearest neighbors only, and used for prediction of a clinical response, $P_{ij}(m,k)$, for the patient j. In our previous work [15], we used the SVM method; in the current work we have tried the kNN method [16], random forest (RF) [17], Tikhonov (ridge) regression (RR) [18], binomial naïve Bayesian method (BNB) [19–22], adaptive boosting (ADA) [7, 23, 24], and multi-layer perceptron (MLP) [25–27]. Repeating this procedure for all other $j \neq i$, we obtain the area-under the ROC curve, $AUC_i(m,k)$, for all but i-th patients for fixed values of data trimming parameters, m and k.

The $AUC_i(m,k)$ value can be then analyzed as a function of data trimming parameters, m and k [15]. Over the range of possible m and k values, we compare the AUC_i function [15]. All pairs of (m,k) values that provide $AUC_i(m,k) > p \cdot \max(AUC_i(m,k))$

constitute the *prediction-accountable set*, S_i, for the patient of interest i [15]. Our experience suggests that reasonable values for the *confidence threshold*, p, are 0.90 or 0.95 [15].

Finally, the FloWPS prediction, P_{Fi}, for the patient of interest i, is calculated via averaging the ML predictions over the prediction-accountable set S_i: $P_{Fi} = mean_{Si}(P_i(m, k))$ [15]. Repeating this procedure for all other patients, we obtain a set of FloWPS predictions for the whole dataset.

2.2 Performance Assessment for Different ML Methods on 21 Cancer Datasets

Note that for any ML methods, after the application of data trimming, the FloWPS predictions, P_{Fi}, are expressed in regression-like terms, i.e. they are likelihoods for attribution of samples to any of two classes (clinical responders or non-responders).

The discrimination threshold (τ), which the user may apply to distinguish between two classes, should be determined according to the cost balance between false positive (FP) and false negative (FN) errors. Generally, the penalty value $P = B \cdot FP + FN$ should be minimized; here B is called relative balance factor. This relative balance factor $B < 1$ for the situations when the FN error (refusal of prescription of a drug, which might help the patient) is more dangerous than the FP one (prescription of a useless, however, not a harmless, treatment). Contrary, $B > 1$, when it is safer not to give a doubtful treatment for a patient then to give it to him/her. Several practitioners of diagnostics tests have different opinions on how high/low should be this balance factor. According to different sources, the preferred values are $B = 4$ [28–30], $B < 0.16$ [31], $4.5 < B < 5$ [32], $B < 5$ [33], $B > 10$ for emergency medicine only [34], $B > 5$ for toxicology [35].

For such a dangerous disease like cancer, B should be lower when few, perhaps only one, treatment option is approved for a certain cancer morphological type and localization, when the refusal to give the unique radical treatment would inevitably doom the patient to death. Contrary, when multiple treatment options are possible, and the doctor has to choose the best one, the risk of prescription of wrong drug should be taken into account, and B should be higher.

We have applied our FloWPS data trimming for 21 cancer datasets. These datasets embraced totally 1778 patients; the biggest dataset had 235 samples, and the smallest one had 41. Among them, ten (10) had expression profiles for breast cancer patients: GSE25066 [36, 37], GSE41998 [38], GSE18728 [39], GSE20181 [40, 41], GSE20194 [42], GSE23988 [43], GSE32646 [44], GSE37946 [45], GSE42822 [46], and GSE59515 [47]. Four (4) datasets represented multiple myeloma: GSE9782 [11], GSE39753 [48], GSE68871 [49] and GSE55145 [50]. Among those for leukemia, one (1) described adult AML cases (GSE5122 [51]), two (2) stayed for pediatric AML [52], and the rest one (1) accounted for pediatric ALL [52, 53]. Additionally, there were one (1) dataset for pediatric Wilms kidney tumor [52], one (1) of low grade glioma [54] and one (1) for lung cancer [54]. Chemotherapeutics included taxanes, bortezomib, vincristine, trastuzumab, letrozole, tipifarnib, temozolomide, busulfan and cyclophosphamide. Among seven ML methods (SVM, kNN, RF, RR, BNB, ADA and MLP) four (SVM, RF, BNB, and MLP) showed the most robust results according to sensitivity and specificity (Fig. 1). Each ML method was applied without and with data trimming. Although different values of relative balance factor B, and, therefore, discrimination threshold τ, do not affect the ROC AUC characteristics (Fig. 1A–D), they are certainly critical for sensitivity (SN, Fig. 1E–H) and specificity (SP, Fig. 1I–L).

The best results for SVM were obtained using the linear kernel with penalty parameter $C = 1$ (Fig. 1A, E, I). For RF, the best results were exhibited under the following parameter settings: $n\_estimators = 30$, $criterion =$ 'entropy' (Fig. 1B, F, J). For BNB, the best parameters were $alpha = 1.0$, $binarize = 0.0$, and $fit\_prior =$ False (Fig. 1C, G, K). For MLP, the best settings were $hidden\_layer\_sizes = 30$, $alpha = 0.001$ (Fig. 1D, H, L). Among these four ML methods, the best results were shown by BNB.

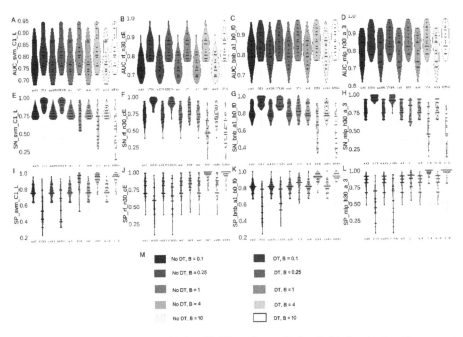

Fig. 1. ROC AUC (A–D), sensitivity (E–H) and specificity (I–L) for treatment response classifiers for eleven newly processed cancer datasets. The classifiers were based on linear SVM with $C = 1$ (A, E, I), as well as RF (B, F, J), BNB (C, G, K) and MLP (D, H, L) with the best parameter settings. The color legend (M) shows absence (No DT) or presence (DT) of flexible data trimming and the value of relative balance factor, B. Within each panel, each *violin* shows distribution of values for 21 cancer datasets: each point stays for a dataset.

3 Discussion

Many ML methods, which are designed for global separation of different classes of points in the feature, space suffer from overtraining, especially in the situations when the number of preceding cases is limited. The global methods may also fail if there is only local rather than global order in placement of different classes in the feature space.

To improve the performance of such global ML methods, we have introduced the elements of local approach, using the flexible data trimming that first avoids extrapolation in the feature space for each validation point, and then selects only several nearest to each validation point neighbors from the training dataset. Note that according to such a hybrid, both global and local simultaneously, approach, for each validation point, training of ML models is done in the specially tailored, individual feature space.

Our approach has confirmed its efficiency for global ML methods, such as SVM, RF, BNB, ADA and MLP. For all 21 cancer gene expression datasets, the use of data trimming with these five ML methods has increased the ROC AUC metric for the classifiers of clinical response on chemotherapy treatment. Moreover, for all these global ML methods with data trimming, 21 out of 21 had ROC AUC metric above the generally acceptable quality threshold (0.70), whereas without data trimming the AUC value exceeded 0.70 not for all these datasets.

Contrary, our data trimming was unable to increase the performance of purely local ML methods such as kNN and RR (data not shown). However, these two local methods suffered from the following disadvantage. Note that regression-like methods, as well as our data trimming, produce as output the continuous value for likelihood of belonging to any class. Contrary, four global ML methods, such as SVM, RF, BNB and MLP, all with data trimming, exhibited more or less acceptable SN and SP in the range for the type I and II error balance factor $0.25 \leq B \leq 4$ (Fig. 1).

Based on a large-scale trial with 21 high throughput clinically annotated gene expression datasets, the BNB method might be advised for further development and implementation for expression-based classifier of individual clinical response to anti-cancer chemotherapy.

4   Methods

All calculations were done using the R package flowpspkg.tar.gz. The package and a manual for it are available at Gitlab through the link: https://gitlab.com/borisov_oncobox/flowpspkg.

Acknowledgements. The study was supported by Russian Foundation for Basic Research Grant 19-29-01108.

References

1. Buzdin, A., et al.: RNA sequencing for research and diagnostics in clinical oncology. Semin. Cancer Biol. (2019). https://doi.org/10.1016/j.semcancer.2019.07.010
2. Zhukov, N.V., Tjulandin, S.A.: Targeted therapy in the treatment of solid tumors: practice contradicts theory. Biochem. Biokhim. **73**, 605–618 (2008)
3. Borisov, N., Buzdin, A.: New paradigm of machine learning (ML) in personalized oncology: data trimming for squeezing more biomarkers from clinical datasets. Front. Oncol. **9**, 658 (2019). https://doi.org/10.3389/fonc.2019.00658

4. Artemov, A., et al.: A method for predicting target drug efficiency in cancer based on the analysis of signaling pathway activation. Oncotarget **6**, 29347–29356 (2015). https://doi.org/10.18632/oncotarget.5119

5. Shepelin, D., et al.: Molecular pathway activation features linked with transition from normal skin to primary and metastatic melanomas in human. Oncotarget **7**, 656–670 (2016). https://doi.org/10.18632/oncotarget.6394

6. Zolotovskaia, M.A., et al.: Pathway based analysis of mutation data is efficient for scoring target cancer drugs. Front. Pharmacol. **10** (2019). https://doi.org/10.3389/fphar.2019.00001

7. Turki, T., Wang, J.T.L.: Clinical intelligence: new machine learning techniques for predicting clinical drug response. Comput. Biol. Med. **107**, 302–322 (2019). https://doi.org/10.1016/j.compbiomed.2018.12.017

8. Turki, T., Wei, Z.: A link prediction approach to cancer drug sensitivity prediction. BMC Syst. Biol. **11** (2017). https://doi.org/10.1186/s12918-017-0463-8

9. Turki, T., Wei, Z., Wang, J.T.L.: Transfer learning approaches to improve drug sensitivity prediction in multiple myeloma patients. IEEE Access **5**, 7381–7393 (2017). https://doi.org/10.1109/ACCESS.2017.2696523

10. Turki, T., Wei, Z., Wang, J.T.L.: A transfer learning approach via procrustes analysis and mean shift for cancer drug sensitivity prediction. J. Bioinform. Comput. Biol. **16**, 1840014 (2018). https://doi.org/10.1142/S0219720018400140

11. Mulligan, G., et al.: Gene expression profiling and correlation with outcome in clinical trials of the proteasome inhibitor bortezomib. Blood **109**, 3177–3188 (2007). https://doi.org/10.1182/blood-2006-09-044974

12. Borisov, N., Tkachev, V., Muchnik, I., Buzdin, A.: Individual Drug Treatment Prediction in Oncology Based on Machine Learning Using Cell Culture Gene Expression Data (2017). https://doi.org/10.1145/3155077.3155078

13. Borisov, N., Tkachev, V., Suntsova, M., Kovalchuk, O., Zhavoronkov, A., Muchnik, I., Buzdin, A.: A method of gene expression data transfer from cell lines to cancer patients for machine-learning prediction of drug efficiency. Cell Cycle **17**, 486–491 (2018). https://doi.org/10.1080/15384101.2017.1417706

14. Borisov, N., Tkachev, V., Buzdin, A., Muchnik, I.: Prediction of drug efficiency by transferring gene expression data from cell lines to cancer patients. In: Rozonoer, L., Mirkin, B., Muchnik, I. (eds.) Braverman Readings in Machine Learning. Key Ideas from Inception to Current State. LNCS (LNAI), vol. 11100, pp. 201–212. Springer, Cham (2018). https://doi.org/10.1007/978-3-319-99492-5_9

15. Tkachev, V., et al.: FLOating-window projective separator (FloWPS): a data trimming tool for support vector machines (SVM) to improve robustness of the classifier. Front. Genet. **9** (2019). https://doi.org/10.3389/fgene.2018.00717

16. Altman, N.S.: An introduction to kernel and nearest-neighbor nonparametric regression. Am. Stat. **46**, 175–185 (1992). https://doi.org/10.1080/00031305.1992.10475879

17. Toloşi, L., Lengauer, T.: Classification with correlated features: unreliability of feature ranking and solutions. Bioinformatics **27**, 1986–1994 (2011). https://doi.org/10.1093/bioinformatics/btr300

18. Tikhonov, A.N., Arsenin, V.I.: Solutions of Ill-Posed Problems. Winston ; Distributed solely by Halsted Press, Washington (1977)

19. Cho, H.-J., Lee, S., Ji, Y.G., Lee, D.H.: Association of specific gene mutations derived from machine learning with survival in lung adenocarcinoma. PLoS ONE **13**, e0207204 (2018). https://doi.org/10.1371/journal.pone.0207204

20. Davoudi, A., Ozrazgat-Baslanti, T., Ebadi, A., Bursian, A.C., Bihorac, A., Rashidi, P.: Delirium prediction using machine learning models on predictive electronic health records data. In: 2017 IEEE 17th International Conference on Bioinformatics and Bioengineering (BIBE), pp. 568–573. IEEE, Washington, DC (2017). https://doi.org/10.1109/BIBE.2017.00014

21. Turki, T., Wei, Z.: Learning approaches to improve prediction of drug sensitivity in breast cancer patients. In: 2016 38th Annual International Conference of the IEEE Engineering in Medicine and Biology Society (EMBC), pp. 3314–3320. IEEE, Orlando, FL, USA (2016). https://doi.org/10.1109/EMBC.2016.7591437

22. Zhang, L., et al.: Applications of machine learning methods in drug toxicity prediction. Curr. Top. Med. Chem. **18** (2018). https://doi.org/10.2174/1568026618666180727152557

23. Wang, Z., et al.: In silico prediction of blood-brain barrier permeability of compounds by machine learning and resampling methods. Chem. Med. Chem. **13**, 2189–2201 (2018). https://doi.org/10.1002/cmdc.201800533

24. Yosipof, A., Guedes, R.C., García-Sosa, A.T.: Data mining and machine learning models for predicting drug likeness and their disease or organ category. Front. Chem. **6** (2018). https://doi.org/10.3389/fchem.2018.00162

25. Minsky, M.L., Papert, S.A.: Perceptrons - Expanded Edition: An Introduction to Computational Geometry. MIT press, Boston (1987)

26. Prados, J., Kalousis, A., Sanchez, J.-C., Allard, L., Carrette, O., Hilario, M.: Mining mass spectra for diagnosis and biomarker discovery of cerebral accidents. Proteomics **4**, 2320–2332 (2004). https://doi.org/10.1002/pmic.200400857

27. Robin, X., Turck, N., Hainard, A., Lisacek, F., Sanchez, J.-C., Müller, M.: Bioinformatics for protein biomarker panel classification: what is needed to bring biomarker panels into *in vitro* diagnostics? Expert Rev. Proteomics **6**, 675–689 (2009). https://doi.org/10.1586/epr.09.83

28. Gent, D.H., Esker, P.D., Kriss, A.B.: Statistical power in plant pathology research. Phytopathology **108**, 15–22 (2018). https://doi.org/10.1094/PHYTO-03-17-0098-LE

29. Ioannidis, J.P.A., Hozo, I., Djulbegovic, B.: Optimal type I and type II error pairs when the available sample size is fixed. J. Clin. Epidemiol. **66**, 903–910.e2 (2013). https://doi.org/10.1016/j.jclinepi.2013.03.002

30. Wetterslev, J., Jakobsen, J.C., Gluud, C.: Trial sequential analysis in systematic reviews with meta-analysis. BMC Med. Res. Methodol. **17**, 39 (2017). https://doi.org/10.1186/s12874-017-0315-7

31. Kim, H.-Y.: Statistical notes for clinical researchers: Type I and type II errors in statistical decision. Restorative Dent. Endodontics **40**, 249 (2015). https://doi.org/10.5395/rde.2015.40.3.249

32. Lu, J., Qiu, Y., Deng, A.: A note on type S/M errors in hypothesis testing. Br. J. Math. Stat. Psychol. **72**, 1–17 (2019). https://doi.org/10.1111/bmsp.12132

33. Litière, S., Alonso, A., Molenberghs, G.: Type I and Type II error under random-effects misspecification in generalized linear mixed models. Biometrics **63**, 1038–1044 (2007). https://doi.org/10.1111/j.1541-0420.2007.00782.x

34. Cummins, R.O., Hazinski, M.F.: Guidelines based on fear of type II (false-negative) errors: why we dropped the pulse check for lay rescuers. Circulation **102**, I377–I379 (2000)

35. Rodriguez, P., Maestre, Z., Martinez-Madrid, M., Reynoldson, T.B.: Evaluating the type II error rate in a sediment toxicity classification using the reference condition approach. Aquat. Toxicol. **101**, 207–213 (2011). https://doi.org/10.1016/j.aquatox.2010.09.020

36. Hatzis, C., et al.: A genomic predictor of response and survival following taxane-anthracycline chemotherapy for invasive breast cancer. JAMA **305**, 1873–1881 (2011). https://doi.org/10.1001/jama.2011.593

37. Itoh, M., et al.: Estrogen receptor (ER) mRNA expression and molecular subtype distribution in ER-negative/progesterone receptor-positive breast cancers. Breast Cancer Res. Treat. **143**, 403–409 (2014). https://doi.org/10.1007/s10549-013-2763-z

38. Horak, C.E., et al.: Biomarker analysis of neoadjuvant doxorubicin/cyclophosphamide followed by ixabepilone or Paclitaxel in early-stage breast cancer. Clin. Cancer Res. **19**, 1587–1595 (2013). https://doi.org/10.1158/1078-0432.CCR-12-1359

39. Korde, L.A., et al.: Gene expression pathway analysis to predict response to neoadjuvant docetaxel and capecitabine for breast cancer. Breast Cancer Res. Treat. **119**, 685–699 (2010). https://doi.org/10.1007/s10549-009-0651-3

40. Miller, W.R., Larionov, A.: Changes in expression of oestrogen regulated and proliferation genes with neoadjuvant treatment highlight heterogeneity of clinical resistance to the aromatase inhibitor, letrozole. Breast Cancer Res. **12**, R52 (2010). https://doi.org/10.1186/bcr2611

41. Miller, W.R., Larionov, A., Anderson, T.J., Evans, D.B., Dixon, J.M.: Sequential changes in gene expression profiles in breast cancers during treatment with the aromatase inhibitor, letrozole. Pharmacogenomics J. **12**, 10–21 (2012). https://doi.org/10.1038/tpj.2010.67

42. Popovici, V., et al.: Effect of training-sample size and classification difficulty on the accuracy of genomic predictors. Breast Cancer Res. **12**, R5 (2010). https://doi.org/10.1186/bcr2468

43. Iwamoto, T., et al.: Gene pathways associated with prognosis and chemotherapy sensitivity in molecular subtypes of breast cancer. J. Nat. Cancer Inst. **103**, 264–272 (2011). https://doi.org/10.1093/jnci/djq524

44. Miyake, T., et al.: GSTP1 expression predicts poor pathological complete response to neoadjuvant chemotherapy in ER-negative breast cancer. Cancer Sci. **103**, 913–920 (2012). https://doi.org/10.1111/j.1349-7006.2012.02231.x

45. Liu, J.C., et al.: Seventeen-gene signature from enriched Her2/Neu mammary tumor-initiating cells predicts clinical outcome for human HER2+: ERα- breast cancer. Proc. Natl. Acad. Sci. U.S.A. **109**, 5832–5837 (2012). https://doi.org/10.1073/pnas.1201105109

46. Shen, K., et al.: Cell line derived multi-gene predictor of pathologic response to neoadjuvant chemotherapy in breast cancer: a validation study on US Oncology 02-103 clinical trial. BMC Med. Genomics **5**, 51 (2012). https://doi.org/10.1186/1755-8794-5-51

47. Turnbull, A.K., et al.: Accurate prediction and validation of response to endocrine therapy in breast cancer. J. Clin. Oncol. **33**, 2270–2278 (2015). https://doi.org/10.1200/JCO.2014.57.8963

48. Chauhan, D., et al.: A small molecule inhibitor of ubiquitin-specific protease-7 induces apoptosis in multiple myeloma cells and overcomes bortezomib resistance. Cancer Cell **22**, 345–358 (2012). https://doi.org/10.1016/j.ccr.2012.08.007

49. Terragna, C., et al.: The genetic and genomic background of multiple myeloma patients achieving complete response after induction therapy with bortezomib, thalidomide and dexamethasone (VTD). Oncotarget **7**, 9666–9679 (2016). https://doi.org/10.18632/oncotarget.5718

50. Amin, S.B., et al.: Gene expression profile alone is inadequate in predicting complete response in multiple myeloma. Leukemia **28**, 2229–2234 (2014). https://doi.org/10.1038/leu.2014.140

51. Raponi, M., et al.: Identification of molecular predictors of response in a study of tipifarnib treatment in relapsed and refractory acute myelogenous leukemia. Clin. Cancer Res. **13**, 2254–2260 (2007). https://doi.org/10.1158/1078-0432.CCR-06-2609

52. Goldman, M., et al.: The UCSC cancer genomics browser: update 2015. Nucleic Acids Res. **43**, D812–D817 (2015). https://doi.org/10.1093/nar/gku1073

53. Tricoli, J.V., et al.: Biologic and clinical characteristics of adolescent and young adult cancers: acute lymphoblastic leukemia, colorectal cancer, breast cancer, melanoma, and sarcoma: biology of AYA cancers. Cancer **122**, 1017–1028 (2016). https://doi.org/10.1002/cncr.29871

54. Tomczak, K., Czerwińska, P., Wiznerowicz, M.: The cancer genome atlas (TCGA): an immeasurable source of knowledge. Contemp. Oncol. (Poznan, Poland) **19**, A68–A77 (2015). https://doi.org/10.5114/wo.2014.47136

Spatio-temporal Tumor Modeling and Simulation (STTMS)

Towards Model-Based Characterization of Biomechanical Tumor Growth Phenotypes

Daniel Abler[1,2]([envelope]) [ORCID], Philippe Büchler[2] [ORCID], and Russell C. Rockne[1] [ORCID]

[1] City of Hope, Duarte, CA, USA
{dabler,rrockne}@coh.org
[2] University of Bern, Bern, Switzerland
{daniel.abler,philippe.buechler}@artorg.unibe.ch

Abstract. Gliomas are the most common malignant brain tumors in adults, with Glioblastoma (GBM) being the most agressive subtype. GBM is clinically evaluated with magnetic resonance imaging (MRI) and presents with different growth phenotypes, involving varying degrees of healthy tissue invasion and tumor induced herniation, also known as mass effect. GBM growth in the brain is frequently modeled as a reaction-diffusion process in which varying ratios of diffusion and proliferation coefficients mimic the observed spectrum of growth phenotypes ranging from nodal to diffuse. However, reaction-diffusion models alone are insufficient to explain tumor-induced mass effect on normal peripheral tissues, which is a critical clinical issue.

We propose an analysis method and framework for estimating GBM growth properties (proliferation, invasiveness, displacive potential) from MRI data routinely collected in the clinical management of GBM. This framework accounts for the mass-effect of the growing tumor by assuming a coupling between local tumor-cell density and volumetric expansion of the tissue.

We evaluate the reconstruction workflow on synthetic data that represents a range of realistic growth situations and levels of uncertainty. For most parameter combinations (90%) that correspond to tumors detectable by T1-weighted MRI, target parameters are recovered with a relative error of less than 15%.

Keywords: Mechanically-coupled tumor growth · Inverse problem · Image-based modeling

1 Tumor Mass-Effect in Glioblastoma

Gliomas are the most frequent malignant brain tumors in adults, with Glioblastoma (GBM) being the most malignant subtype. The rapid invasive growth of this tumor frequently results in lesions that cause healthy-tissue deformation, midline shift or herniation. Biomechanical forces, such as those caused by the

© Springer Nature Switzerland AG 2019
G. Bebis et al. (Eds.): ISMCO 2019, LNCS 11826, pp. 75–86, 2019.
https://doi.org/10.1007/978-3-030-35210-3_6

growing tumor, are known to shape the tumor environment and contribute to tumor progression [6]. Additionally, in brain tumors, elevated solid stress is linked to neuronal loss and neurological dysfunction [10]. In GBM patients, increased tumor mass-effect has been shown to be associated to poor prognosis [11]. This suggests that the propensity of an individual tumor to displace healthy tissue can provide information about the tumor micro-environment and might be of predictive value for treatment and outcome. However, tumors of similar imaging volumes have been observed to give rise to different amounts of tumor mass-effect [11], Fig. 1, possibly resulting in distinct mechanical stress distributions and magnitudes.

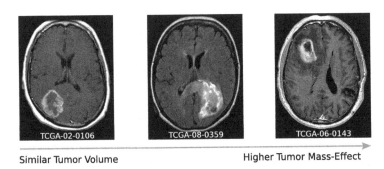

Fig. 1. Tumors of similar imaging volume can exhibit different degrees of mass-effect. Images from TCGA-GBM.

The growth characteristics of GBM have been studied extensively using mathematical models that describe the invasive growth of this tumor as a reaction-diffusion process. These models characterize GBM growth phenotypes on a spectrum from mostly *nodular* to mostly *diffuse*, corresponding to *invasive* growth. However, it remains unknown whether differences in proliferative and invasive potential are sufficient to explain the observed differences in mass-effect.

To investigate the relation between *proliferation, invasiveness*, tumor mass-effect and its manifestation on clinical imaging, we are developing a framework for characterizing *mechanically-coupled* GBM growth. By finding solutions to the inverse growth problem, we aim to establish whether *proliferation* and *invasiveness* can explain the observed variability in tumor mass-effect, or whether distinct biomechanical growth phenotypes of GBM exist that differ also in their *"displaciveness"*.

Here, we present an approach for estimating parameters of a mathematical tumor growth model that accounts for the mass effect of the tumor. We propose a workflow for applying this approach to MR imaging data, and evaluate its accuracy and robustness in a parametric study on 2D synthetic data that represents a range of realistic growth situations and levels of uncertainty.

2   Materials and Methods

2.1   Mathematical Model of Mechanically-Coupled Tumor Growth

Mathematical models of tumor mass-effect were initially studied in the context of atlas-based image segmentation [7]. These models were soon extended to account for tumor growth dynamics by coupling to single-species reaction diffusion equations [3]. More recently, information about tumor induced mechanical-stresses has been incorporated in biophysical tumor growth models to inform local motility of tumor cells in the brain [5], and multi-species mechanically-coupled growth models have been developed [12].

Here we use a single-species mechanically-coupled reaction-diffusion model [1] that captures the dominant aspects of macroscopic GBM growth: the diffuse invasion of the growing tumor into surrounding healthy tissue, and the resulting mass effect.

Invasive growth is modeled phenomenologically as a reaction-diffusion process:

$$\frac{\partial c}{\partial t} = \nabla \cdot (D \nabla c) + \rho c \left(1 - c\right) c, \tag{1}$$

with normalized cancer cell density $c\left(\boldsymbol{x}, t\right)$ and diffusion coefficient $D = D(\boldsymbol{x})$. Tumor cell proliferation is assumed to follow logistic growth with proliferation rate $\rho = \rho\left(\boldsymbol{x}\right)$.

To simulate the tissue-displacing mass-effect of the growing tumor, we model the growth domain as elastic continuum in which the actual deformation $\boldsymbol{u}\left(\boldsymbol{x}, t\right)$ of a tissue element is given by the combination of growth-induced strains $\hat{\boldsymbol{\epsilon}}^{\text{growth}}$ and strains caused by the elastic response of the tissue. We assume a linear constitutive relation between mechanical stress $\hat{\boldsymbol{\sigma}}$ and strain $\hat{\boldsymbol{\epsilon}}$, as well as mechanically isotropic materials that are fully characterized by Young's modulus E and Poisson ratio ν.

$$\hat{\boldsymbol{\sigma}}(\boldsymbol{u}) = \frac{E}{2\left(1 + \nu\right)} \hat{\epsilon}(\boldsymbol{u}) + \frac{E\,\nu}{\left(1 + \nu\right)\left(1 - 2\nu\right)} \operatorname{Tr} \hat{\epsilon}(\boldsymbol{u})\, \mathbb{1} \tag{2a}$$

$$\hat{\epsilon}(\boldsymbol{u}) = \frac{1}{2} \left(\nabla \boldsymbol{u} + \left(\nabla \boldsymbol{u}\right)^{\mathrm{T}}\right) \tag{2b}$$

Additionally, we postulate a linear coupling between tumor cell density and growth-induced strain with isotropic coupling strength λ:

$$\hat{\epsilon}^{\text{growth}}(c) = \lambda \mathbb{1} c. \tag{3}$$

Table 1 summarizes variables and parameters of this model.

The model is implemented using the *FEniCS* library[1] [2] for solving the model equations via the Finite Element Method. This implementation employs first and second order Lagrange elements for spatial interpolation of displacement $\boldsymbol{u}\left(\boldsymbol{x}\right)$ and density $c\left(\boldsymbol{x}\right)$ fields, respectively. Time-stepping is performed using a first order implicit numerical scheme.

---

[1] https://fenicsproject.org.

Table 1. Variables and parameters of the mathematical model.

Symbol	Parameter name	Units
$c(\boldsymbol{x}, t)$	tumor cell density	normalized to c_0
$\boldsymbol{u}(\boldsymbol{x}, t)$	tumor-induced displacements	mm
D	diffusion coefficient/diffusivity	mm^2/d
ρ	proliferation rate	$1/d$
λ	coupling constant	
E	young's modulus	kPa
ν	poisson's ratio	

2.2   Simulation Domain

Growth is simulated in a 2D computational domain Ω based on the $SRI24$[2] [8] atlas of normal human brain anatomy. The atlas contains tissue classes for White Matter (WM), Grey Matter (GM) and Cerebrospinal Fluid (CSF). The latter was divided into two compartments to distinguish fluid-filled brain ventricles from the remaining CSF, Fig. 2. Distinct isotropic growth and mechanical tissue parameters D_i, ρ_i, E_i, ν_i were assigned to each subdomain Ω_i.

The simulation domain was spatially discretized into a mesh of triangular elements with maximum cell diameter of 1.42 mm. We assumed the growth domain to be free of any initial mechanical stresses and approximate the displacement constraint imposed by the rigid skull by zero-displacement Dirichlet boundary conditions on the domain boundary. Similarly, tumor cells were prevented from leaving the domain by zero-flux von-Neumann boundary conditions. The tumor was initialized by a Gaussian-shaped 2D tumor cell density field $c_0 = c(\boldsymbol{x}_0, t = 0)$ centered at the seed location $\boldsymbol{x}_0$ and with standard-deviation of 1 mm.

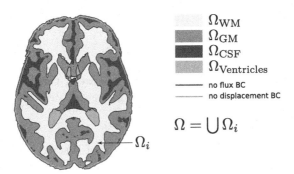

Fig. 2. Schema of brain simulation domain Ω with subdomains Ω_i for white matter (WM), grey matter (GM), surrounding cerebrospinal fluid (CSF) and CSF-filled ventricles.

---

[2] https://www.nitrc.org/projects/sri24/.

2.3 Estimation of Growth Parameters as Inverse Problem

Given the forward model, Eqs. (1) to (3), growth parameter identification can be framed as a PDE-constrained optimization problem with the aim to identify the set of parameters p^{opt} that minimizes an objective functional J. Such a functional can be constructed by comparing the primary variables of the simulation, $c(x)$ and $u(x)$ to corresponding target tumor cell density $c^*(x)$ and tissue displacement $u^*(x)$ fields at a specific observation time point k.

$$J = \|c(x, t_k) - c^*(x, t_k)\|_2^2 + \|u(x, t_k) - u^*(x, t_k)\|_2^2 \qquad (4)$$

Simulated tumor cell density $c(x, t_k)$ and tissue deformation $u(x, t_k)$ fields at the corresponding simulation time step t_k are constrained by the forward model and depend on the current set of simulation parameters p.

While medical imaging provides information about those quantities, detailed spatial maps are not directly observable. A commonly used approached in the GBM modeling literature estimates tumor cell density by associating specific imaging detection thresholds to different imaging modalities. Tumor features visible in T1- and T2-weighted MR-imaging have been linked to different levels of relative tumor cell density: $c > 0.80$ for visibility on T1-weighted contrast enhanced MRI, and $c > 0.16$ for visibility on T2-weighted (T2) MRI [14]. Routine clinical imaging for brain tumors thus provides two views of the unknown tumor cell density field $c^*(x)$, corresponding to two indicator functions $\chi_{T1}^*(x)$ and $\chi_{T2}^*(x)$ that identify the positions x where $c(x) \geq 0.80$ and $c(x) \geq 0.16$, respectively. Tissue displacements in the brain can be estimated by deformable image registration between two imaging time points or relative to a healthy brain atlas, which allows an estimate for the tumor-induced displacement field $\tilde{u}^*(x)$ to be obtained at diagnosis and between follow-up scans.

Given these target fields, an alternative objective function based on image-derivable target quantities can be formulated:

$$J = \|\chi_{T1}(x, t_k) - \chi_{T1}^*(x, t_k)\|_2^2 + \|\chi_{T2}(x, t_k) - \chi_{T2}^*(x, t_k)\|_2^2$$
$$+ \|u(x, t_k) - \tilde{u}^*(x, t_k)\|_2^2 \qquad (5)$$

with $\chi_i(x, t_k)$ obtained by applying the respective detection threshold to the simulated density field $c(x, t_k)$.

The adjoint method provides an efficient approach for computing the gradient $\frac{dJ}{dp}$ and thus for solving the minimization problem $\min_p (J)$. This implementation uses the *dolfin-adjoint* library[3] for automatic derivation of the discrete adjoint equations for our forward model, Eqs. (1) to (3), and optimization functionals Eqs. (4) and (5), respectively.

2.4 Evaluation of Parameter Estimation Approach

We evaluated the performance of this parameter estimation approach in two different scenarios using synthetic data generated from simulation of the for-

---

[3] http://www.dolfin-adjoint.org.

Table 2. Parameter ranges for parametric study. Growth parameters D_{WM}, ρ, λ were varied across physiological ranges (min, max) resulting in 100 parameter combinations. A fixed relation was assumed between diffusivity in GM and WM: $D_{WM} = 5 \cdot D_{GM}$ [13].

Parameter	min	max	step	# steps	init	Units
D_{WM}	0.05	0.20	0.05	4	0.001	mm^2/d
ρ	0.02	0.18	0.04	5	0.001	$1/d$
λ	0.02	0.18	0.04	5	0.001	

ward model. In both cases, the duration of tumor growth T, as well as initial conditions, tumor seed location and zero initial displacements, were assumed to be known. Mechanical tissue properties were fixed to $E_{WM/GM} = 3.00\,kPa$, $E_{CSF} = 1.00\,kPa$, $\nu_{WM/GM} = 0.45$, $\nu_{CSF} = 0.30$.

Reconstruction from Forward Simulation: First, we aimed to recover the simulation parameters of the forward model $p = \{D_{WM}, D_{GM}, \rho_{WM}, \rho_{GM}, \lambda\}$, directly from results of the forward simulation, using density $c(x, T)$ and displacement $u(x, T)$ fields from the final simulation time point T as reference, Eq. (4).

Reconstruction from Image-Derived Target Fields: Second, we studied a more realistic scenario in which we aimed to recover the simulation parameters from information available from routine clinical MR imaging, Eq. (5). This scenario accounts for the noise associated to the derivation of target fields χ^*_{T1}, χ^*_{T2}, u^* from this information. In this setting, we characterized the performance of the proposed parameter estimation approach in a parametric study by sampling $(n = 100)$ from realistic ranges of three independent growth parameters $p = \{D_{WM}, \rho, \lambda\}$, Table 2.

For each parameter combination p, tumor growth was simulated for a time period T with time steps $\Delta t = 1d$, Fig. 3(A). At the final simulation time point, density $c(x, T)$ and displacement $u(x, T)$ fields were extracted and used to construct a synthetic dataset that mimics the kind of information that can be obtained from routine clinical MR imaging, Fig. (3)(B). The simulated density field $c(x, T)$ was subjected to thresholds $c(x) \geq 0.16$ and $c(x) \geq 0.80$, resulting in two indicator functions corresponding to the portion of the tumor visible on T2-weighted and T1-weighted MRI. We used these indicator functions χ^*_{T1}, χ^*_{T2} as target fields in the optimization process, Eq. (5). The simulated displacement field $u(x, T)$ was used to deform the anatomical (T1 MRI) atlas on which growth had been simulated. From the resulting images we estimated the tumor-induced displacement by deformable image registration, using the symmetric image normalization method (SyN) as implemented in the *Advanced Normalization Tools* (ANTs)[4]. This reconstructed displacement field $\tilde{u}^*$ served as target field in the optimization process, Fig. (3)(C).

---

[4] https://github.com/ANTsX/ANTs.

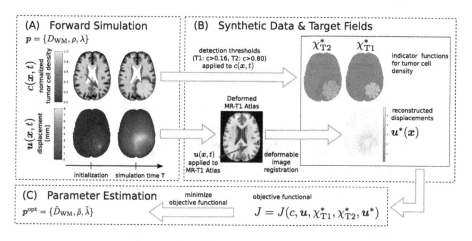

Fig. 3. Workflow of parametric study: Results from forward simulation (A) were used to create synthetic target fields (B) to which the proposed parameter estimation approach (C) was applied to recover the original simulation parameters $p^{\mathrm{opt}} \approx p$.

We applied this approach to each synthetic data set to obtain a set of *reconstructed* growth parameters $p^{\mathrm{opt}} = \{\tilde{D}_{\mathrm{WM}}, \tilde{\rho}, \tilde{\lambda}\}$. Parameter optimization was initialized with the values indicated in column *init* in Table 2. Duration of tumor growth T and initial conditions were assumed to be known in each optimization scenario.

We compared reconstructed p^{opt} to actual p growth parameters in terms of their absolute value and relative reconstruction error $\epsilon_i = (p_i^{\mathrm{opt}} - p_i)/p_i$.

3   Results

3.1   Forward Simulation

Tumor evolution and tumor-induced mass-effect were simulated for $T = 250\,\mathrm{d}$ days forward in time starting from an initial Gaussian-shaped tumor cell distribution. Figure 4 illustrates the evolution of tumor cell density $c(x, T)$ and the resulting tumor-induced displacement field $u(x, T)$. Compression of the lateral ventricles by the growing tumor is evident from the last row of Fig. 4.

3.2   Reconstruction from Forward Simulation

Forward simulation over $T = 250\,\mathrm{d}$ was repeated for seed positions in three different locations, indicated by red arrows in Fig. 5: in GM (*Case 1*), WM (*Case 2*), and at the interface between GM and WM (*Case 3*). Using density and displacement fields from the final time point, $c^*(x, T)$, $u^*(x, T)$, we tried to recover the simulation parameters of the forward model by PDE-constrained optimization.

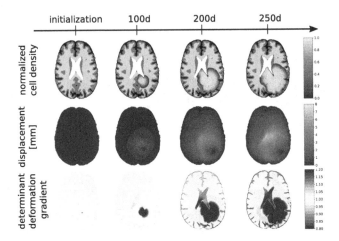

Fig. 4. Simulated evolution of tumor growth and mass effect.

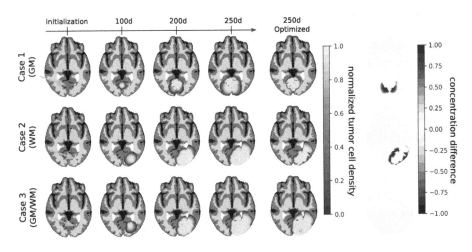

Fig. 5. Simulated evolution of tumor growth from seeds in three different locations: GM (*Case 1*), WM (*Case 2*), and at the interface between GM and WM (*Case 3*). Simulated density fields based on the reference parameters p^{ref} are compared ($T = 250\text{d}$) to density fields based on the parametersets obtained from optimization (p^{opt1}, p^{opt2}, p^{opt3}, see Table 3). Note that GBM very rarely grow or migrate into the cerebellum. The seed locations have been chosen to illustrate the parameter estimation approach on approximately equally sized contiguous patches of GM and WM. (Color figure online)

Table 3 summarizes reference parameters for the forward simulation p^{ref}, their initialization for optimization p^{init} and optimization results p^{opt} for the three scenarios depicted in Fig. 5.

Reference growth parameters could be recovered correctly for the brain region most affected by the tumor: GM properties for *Case 1* and WM properties for *Case 2*. For *Case 3*, which grew with substantial involvement of both WM and GM domains, target parameters for both regions were recovered correctly.

Table 3. Reference parameters for forward simulations and reconstructed parameters for cases 1 to 3 in Fig. 5.

	ρ_{WM} [1/d]	ρ_{GM} [1/d]	D_{WM} [mm^2/d]	D_{GM} [mm^2/d]	λ
Forward model reference p^{ref}	0.08	0.080	0.100	0.020	0.150
Optimization initialization p^{init}	0.010	0.010	0.010	0.010	0.200
Case 1 p^{opt1}	0.010	0.080	0.010	0.020	0.150
Case 2 p^{opt2}	0.080	0.010	0.100	0.010	0.150
Case 3 p^{opt3}	0.080	0.080	0.100	0.020	0.150

3.3   Reconstruction from Image-Derived Target Fields

In a second evaluation, the workflow shown in Fig. 3 was applied to $n = 100$ combinations of growth parameters $p = \{D_{\text{WM}}, \rho, \lambda\}$, Table 2, for a duration of $T = 100$ d in two different growth domains.

Figure 6 compares the resulting distribution of reconstructed parameter values to their target values. Distributions of reconstructed D_{WM} and ρ_{WM} are concentrated around the respective target values Estimates corresponding to the highest parameter values explored in this study ($D_{\text{WM}} = 0.20$ mm^2/d, $\rho_{\text{WM}} = 0.18$ d^{-1}) show the largest uncertainty with few outliers extending their distribution towards values below the target value. Most reconstructions of λ slightly overestimate the target value; parameter estimates for the lowest coupling ($\lambda = 0.02$) are associated with highest uncertainty and biased towards larger values.

Only tumors that are sufficiently large and dense to be detected on both T1 and T2 weighted MRI were included in Fig. 6. Slowly or diffusively growing tumors with high D/ρ ratios may not be visible in one or both MR modalities and therefore do not contribute to the objective functional, Eq. (5), resulting in higher relative reconstruction errors for this group of parameter combinations. Figure 7 shows the fraction of reconstructed parameter sets in function of the maximum relative reconstruction error in each parameter, and visibility of the tumor on

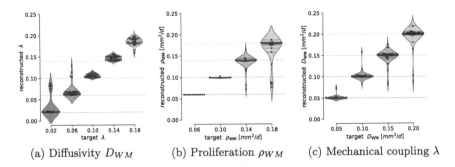

(a) Diffusivity D_{WM} (b) Proliferation ρ_{WM} (c) Mechanical coupling λ

Fig. 6. Distribution of reconstructed parameter values versus target values. Horizontal lines indicate the target value used in the forward simulation. Only simulations with $c(x, T)$ exceeding T1 and T2 detection thresholds are included; therefore, no reconstructed values are reported for simulation parameter $\rho = 0.02d^{-1}$ in (b).

T1- and T2-weighted MR imaging. Provided that the tumor is detectable in T1-weighted MRI, we obtained relative reconstruction errors of less than 15 % for about 90 % of all converged optimization cases (190 of 200) across the parameter space.

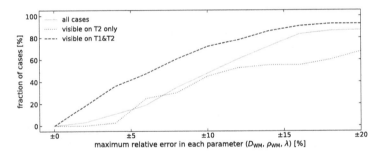

Fig. 7. Fraction of reconstructed parameter sets in function of the maximum relative reconstruction error. A value of $\pm\zeta$ on the x-axis indicates that all relative reconstruction errors ϵ_i simultaneously fulfill $|\epsilon_i| < |\zeta|$ for all parameters $i \in \{D_{\mathrm{WM}}, \rho_{\mathrm{WM}}, \lambda\}$.

4   Discussion

This study proposes an imaging biomarker for tumor mass-effect that can be derived from parametrizations of a biomechanically-coupled tumor growth model.

While adjoint-based optimization of parameters in a similar mechanically coupled tumor growth model had been explored before [4], the characterization of this approach was limited to 1D, and tumor-induced tissue deformation was taken into account only at selected landmark positions. More recent advances include frameworks, such as [9] which combines adjoint parameter estimation of a reaction diffusion model with image registration, but does not explicitly model tumor mass-effect.

Here, we presented a method for estimating parameters of a mechanically coupled tumor growth model from routine clinical imaging information of glioma patients. Performance of this method was characterized on 2D synthetic data in a reconstruction workflow that mimics data and associated uncertainties of real reconstruction scenarios. We demonstrated self-consistency of this approach and found relative reconstruction errors of less than 15 % for about 90 % of cases, provided that the tumor is detectable in T1 weighted MRI.

This study assumed that origin and duration of tumor growth were known for the optimization process. However, and particularly when applied to a single observation at the time of diagnosis, tumor origin and duration of growth may be unknown. Preliminary tests indicate similar reconstruction performance when seeding the optimization process at the center-of-mass position of the observed synthetic tumor. Estimated values of the ratio D/ρ are expected to be independent of the growth period. However, the degree to which D/ρ and λ can be identified simultaneously under this condition remains to be studied.

We aim to use the developed framework to characterize GBM growth in terms of the tumor's invasiveness and its displacive potential. As next steps, we plan to characterize this approach in 3D and to investigate its application to patient MR images.

Acknowledgement. The research leading to these results has received funding from the European Union's Horizon 2020 research and innovation programme under the Marie Skłodowska-Curie grant agreement No. 753878.

References

1. Abler, D., Büchler, P.: Evaluation of a mechanically coupled reaction–diffusion model for macroscopic brain tumor growth. In: Gefen, A., Weihs, D. (eds.) Computer Methods in Biomechanics and Biomedical Engineering. LNB, pp. 57–64. Springer, Cham (2018). https://doi.org/10.1007/978-3-319-59764-5_7
2. Alnæs, M., et al.: The FEniCS Project Version 1.5. Archive of Numerical Software, vol. 3 (2015). https://doi.org/10.11588/ans.2015.100.20553
3. Clatz, O., et al.: Realistic simulation of the 3-D growth of brain tumors in MR images coupling diffusion with biomechanical deformation. IEEE Trans. Med. Imaging **24**(10), 1334–1346 (2005). https://doi.org/10.1109/TMI.2005.857217
4. Hogea, C., Davatzikos, C., Biros, G.: An image-driven parameter estimation problem for a reaction-diffusion glioma growth model with mass effects. J. Math. Biol. **56**(6), 793–825 (2008). https://doi.org/10.1007/s00285-007-0139-x
5. Hormuth, D.A., Eldridge, S.L., Weis, J.A., Miga, M.I., Yankeelov, T.E.: Mechanically coupled reaction-diffusion model to predict glioma growth: methodological details. In: von Stechow, L. (ed.) Cancer Systems Biology. MMB, vol. 1711, pp. 225–241. Springer, New York (2018). https://doi.org/10.1007/978-1-4939-7493-1_11

6. Jain, R.K., Martin, J.D., Stylianopoulos, T.: The role of mechanical forces in tumor growth and therapy. Ann. Rev. Biomed. Eng. **16**(1), 321–346 (2014). https://doi.org/10.1146/annurev-bioeng-071813-105259

7. Mohamed, A., Davatzikos, C.: Finite element modeling of brain tumor mass-effect from 3D medical images. In: Duncan, J.S., Gerig, G. (eds.) MICCAI 2005. LNCS, vol. 3749, pp. 400–408. Springer, Heidelberg (2005). https://doi.org/10.1007/11566465_50

8. Rohlfing, T., Zahr, N.M., Sullivan, E.V., Pfefferbaum, A.: The SRI24 multichannel atlas of normal adult human brain structure. Human Brain Mapping **31**(5), 798–819 (2010). https://doi.org/10.1002/hbm.20906

9. Scheufele, K., Mang, A., Gholami, A., Davatzikos, C., Biros, G., Mehl, M.: Coupling brain-tumor biophysical models and diffeomorphic image registration. Comput. Methods Appl. Mech. Eng. **347**, 533–567 (2019)

10. Seano, G., et al.: Solid stress in brain tumours causes neuronal loss and neurological dysfunction and can be reversed by lithium. Nature Biomed. Eng., January 2019. https://doi.org/10.1038/s41551-018-0334-7

11. Steed, T.C., et al.: Quantification of glioblastoma mass effect by lateral ventricle displacement. Sci. Rep. **8**(1), December 2018. https://doi.org/10.1038/s41598-018-21147-w

12. Subramanian, S., Gholami, A., Biros, G.: Simulation of glioblastoma growth using a 3d multispecies tumor model with mass effect. J. Math. Biol. **79**(3), 941–967 (2019). https://doi.org/10.1007/s00285-019-01383-y

13. Swanson, K.R., Alvord, E.C., Murray, J.D.: A quantitative model for differential motility of gliomas in grey and white matter. Cell Prolif. **33**(5), 317–329 (2000)

14. Swanson, K.R., Rostomily, R.C., Alvord, E.C.: A mathematical modelling tool for predicting survival of individual patients following resection of glioblastoma: a proof of principle. Br. J. Cancer **98**(1), 113–119 (2008)

Population Modeling of Tumor Growth Curves, the Reduced Gompertz Model and Prediction of the Age of a Tumor

Cristina Vaghi[1,2], Anne Rodallec[3], Raphaelle Fanciullino[3], Joseph Ciccolini[3], Jonathan Mochel[4], Michalis Mastri[5], John M. L. Ebos[5], Clair Poignard[1,2], and Sebastien Benzekry[1,2(✉)]

[1] MONC Team, Inria Bordeaux Sud-Ouest, Bordeaux, France
sebastien.benzekry@inria.fr
[2] Institut de Mathématiques de Bordeaux, Bordeaux, France
[3] SMARTc, Center for Research on Cancer of Marseille, Marseille, France
[4] Department of Biomedical Sciences, Iowa State University, Ames, USA
[5] Roswell Park Comprehensive Cancer Center, Buffalo, NY, USA

Abstract. Quantitative analysis of tumor growth kinetics has been widely carried out using mathematical models. In the majority of cases, individual or average data were fitted.

Here, we analyzed three classical models (exponential, logistic and Gompertz within the statistical framework of nonlinear mixed-effects modelling, which allowed us to account for inter-animal variability within a population group. We used *in vivo* data of subcutaneously implanted Lewis Lung carcinoma cells. While the exponential and logistic models failed to accurately fit the data, the Gompertz model provided a superior descriptive power. Moreover, we observed a strong correlation between the Gompertz parameters. Combining this observation with rigorous population parameter estimation motivated a simplification of the standard Gompertz model in a reduced Gompertz model, with only one individual parameter. Using Bayesian inference, we further applied the population methodology to predict the individual initiation times of the tumors from only three measurements. Thanks to its simplicity, the reduced Gompertz model exhibited superior predictive power.

The method that we propose here remains to be extended to clinical data, but these results are promising for the personalized estimation of the tumor age given limited data at diagnosis.

Keywords: Tumor growth kinetics · Gompertz model · Mixed-effects modeling · Bayesian estimation

1 Introduction

Tumor growth kinetics have been studied since several decades both clinically [8] and experimentally [18]. One of the findings of these early studies is that

© Springer Nature Switzerland AG 2019
G. Bebis et al. (Eds.): ISMCO 2019, LNCS 11826, pp. 87–97, 2019.
https://doi.org/10.1007/978-3-030-35210-3_7

tumor growth is not exponential provided it is observed on a long enough time frame (100 to 1000 folds of increase) [13]. The specific growth rate slows down and this deceleration can be particularly well captured by the Gompertz model [13,15,21]. The analytical expression of this model writes (where V_0 is the initial tumor size at $t = 0$ and α and β are two parameters):

$$V(t) = V_0 e^{\frac{\alpha}{\beta}\left(1-e^{-\beta t}\right)} \tag{1}$$

While the etiology of the Gompertz model has been long debated [10], several independent researchers have reported a strong correlation between the parameters α and β estimated on distinct subjects within the same species [6,13,16]. While some suggested this would imply a constant maximal tumor size (given by $V_0 e^{\frac{\alpha}{\beta}}$ in (1)) across tumor types within a given species [6], others argued that because of the presence of the exponential, this could vary over several orders of magnitude [19]. To date, the generalizability, implications and understanding of this observation remain open questions in quantitative tumor growth.

Mathematical models for tumor growth have been previously studied at the level of individual kinetics and for prediction of future tumor growth [2]. However, up to our knowledge, a detailed study of statistical properties of classical growth models at the level of the population (*i.e.* integrating structural dynamics with inter-animal variability) remains yet to be reported. Longitudinal data analysis with nonlinear mixed-effect modelling provides an ideal tool for such a task [14]. In addition, the reduced number of parameters (from $p \times N$ to $p + \frac{p(p+1)}{2}$ where N is the number of animals and p the number of parameters of the model) ensures a higher robustness of the estimates, in the sense of smaller standard errors. Therefore, this framework is particularly adapted to study the above-mentioned correlation between the two Gompertz parameters.

Moreover, using the population distribution as prior allows to make predictions on new subjects by means of Bayesian algorithm such as the Hamiltonian Monte Carlo algorithm [11,12], implemented in Stan [7]. The advantage of this method is that only few measurements of the new individual are necessary to have reliable prognosis.

2   Materials and Methods

Mice Experiments. The experimental data consisted in murine Lewis lung carcinoma cells originally derived from a spontaneous tumor in a C57BL/6 mouse [4]. They were implanted subcutaneously (10^6 cells at injection) on the caudal half of the back in anesthetized 6- to 8-week-old C57BL/6 mice. Tumor size was measured as described for the breast data. The data was pooled from two experiments with a total of 188 observations. A precise description of the experimental protocol is reported elsewhere (see [2]).

Tumor Growth Models. At the time of injection ($t_0 = 0$), we assumed that all the animal tumor volumes within a group have the same volume V_0 (taken to be equal to the number of injected cells converted in mm^3) and denote by α the specific growth rate ($\alpha = \frac{1}{V}\frac{dV}{dt}$) at this time and volume.

We considered the exponential, logistic and Gompertz models [2]. The first two are respectively defined by:

$$V_E(t;\alpha) = V_0\exp(\alpha t) \qquad \text{and} \qquad V_L(t;\alpha, K)\frac{V_0 K}{(V_0 + (K - V_0)e^{-\alpha t})}. \qquad (2)$$

In the logistic equation, K is a carrying capacity parameter.

The Gompertz model $V_G(t;\alpha,\beta)$ is characterized by an exponential decrease of the specific growth rate with rate β. The differential form thus reads:

$$\begin{cases} \dfrac{dV_G}{dt} = \left(\alpha - \beta\log\left(\dfrac{V_G}{V_0}\right)\right)V_G, \\ V_G(t = 0) = V_0. \end{cases} \qquad (3)$$

Note here that the initial condition also appears in the differential equation defining V_G. This is natural from our assumption that α is the specific growth rate at the injected volume V_0.

Population Approach. Let N be the total number of subjects within the population and $\boldsymbol{Y}^i = \{y_1^i, \ldots, y_{n^i}^i\}$ the vector of longitudinal measurements of the animal i, where y_j^i is the observation of subject i at time t_j^i for $i = 1, \ldots, N$ and $j = 1, \ldots, n^i$ (n^i is the total number of measurements of individual i). We assumed the following statistical model

$$y_j^i = V(t_j^i; \boldsymbol{\theta}^i) + e_j^i, \qquad j = 1, \ldots, n^i, \quad i = 1, \ldots, N, \qquad (4)$$

where $V(t_j^i; \boldsymbol{\theta}^i)$ is the evaluation of one of the tumor growth models at time t_j^i, $\boldsymbol{\theta}^i \in \mathbb{R}^p$ is the vector of the parameters relative to the individual i and e_j^i the residual error model, to be defined later. We assumed that the individual parameters $\boldsymbol{\theta}^i$ follow a lognormal distribution that are therefore identified by

$$\log(\boldsymbol{\theta}^i) = \log(\boldsymbol{\mu}) + \boldsymbol{\eta}^i,$$

where $\boldsymbol{\mu}$ denotes the fixed effects and $\boldsymbol{\eta}^i$ denotes the random effects. The former are identical within the population while the latter are specific for each animal and follow a normal distribution $\boldsymbol{\eta}^i \sim \mathcal{N}(0, \boldsymbol{\omega})$ with mean zero and variance matrix $\boldsymbol{\omega}$.

We considered a combined residual error model e_j^i, defined as

$$e_j^i = \left(\sigma_1 + \sigma_2 f(t_j^i; \boldsymbol{\theta}^i)\right)\varepsilon_j^i,$$

where $\varepsilon_j^i \sim \mathcal{N}(0,1)$ are the residual errors and (σ_1, σ_2) are the residual error model parameter.

In order to compute the population parameters, we maximized a population likelihood, obtained by pooling together all the data. Usually, this likelihood cannot be computed explicitly for nonlinear mixed-effect models. The optimization procedure can be implemented using the stochastic approximation expectation minimization algorithm (SAEM) [14], implemented in `Monolix` [1].

From now on we denote by $\phi = \{\boldsymbol{\mu}, \boldsymbol{\omega}, \boldsymbol{\sigma}\}$ the set of the population parameters containing the fixed effects $\boldsymbol{\mu}$ and the random effects $\boldsymbol{\omega}$ of the parameters and the vector of error model parameters $\boldsymbol{\sigma} = [\sigma_1, \sigma_2]$.

Individual Predictions: Bayesian Inference. We considered the problem of predicting the age of the tumor of an animal based on three late measurements. We splitted the data set into two subgroups: a *training set*, used to learn the population parameters distribution, and a *test set*, to assess the performance of the prediction.

Let us assume that the set of the population parameters ϕ has been identified on a *training set* using the population approach. We used this information to make predictions for a new animal j in the *test set* considering only its last three measurements $\boldsymbol{y}^j = \{y_{n^j-2}^j, y_{n^j-1}^j, y_{n^j}^j\}$. The posterior distribution $\mathbb{P}(\boldsymbol{\theta}^j | \boldsymbol{y}^j, \phi)$ of the parameters $\boldsymbol{\theta}^j$ was then given thanks to the Bayesian approach [11]:

$$\mathbb{P}(\boldsymbol{\theta}^j | \boldsymbol{y}^j; \phi) = \mathbb{P}(\boldsymbol{\theta}^j; \phi)\mathbb{P}(\boldsymbol{y}^j | \boldsymbol{\theta}^j; \phi), \tag{5}$$

where $\mathbb{P}(\boldsymbol{\theta}^j; \phi)$ is the prior distribution of the parameters found with the nonlinear mixed effects modeling and $\mathbb{P}(\boldsymbol{y}^j | \boldsymbol{\theta}^j; \phi)$ is the likelihood. Then we computed the posterior predictive distribution of $\tilde{y}^j(u)$, with $u < t_{n^j-2}$ defined as

$$\mathbb{P}(\tilde{y}^j(u) | \boldsymbol{y}^j) = \int_{\theta^j} \mathbb{P}(\tilde{y}^j(u) | \boldsymbol{\theta}^j; \phi)\mathbb{P}(\boldsymbol{\theta}^j | \boldsymbol{y}^j; \phi)d\boldsymbol{\theta}^j. \tag{6}$$

We draw realizations for (5) and for (6) using `Pystan`, a Python interface to the software `Stan` [7] for Bayesian inference based on the No-U-Turn sampler, a variant of Hamiltonian Monte Carlo [12]. These realizations were then used to estimate tumor growth kinetic as the median value of the sample.

3   Results

In [20] other two data sets (two animal models of breast cancer, measured by volume and fluorescence) are considered for the analysis with equivalent results.

3.1   Population Analysis of Tumor Growth Curves

We applied the population approach to test the descriptive power of the exponential, logistic and Gompertz models for tumor growth kinetics. The number of injected cells at time $t_0 = 0$ was 10^6, therefore we fixed the initial volume $V_0 = 1\,\text{mm}^3$ in the whole dataset [2].

We ran the SAEM algorithm with the `Monolix` software to estimate the fixed and random effects. Moreover, different statistical indices were evaluated in order to compare the different tumor growth models. We report them in Table 1, where

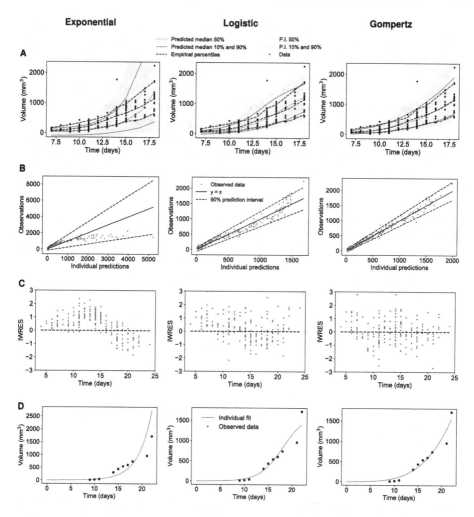

Fig. 1. Results relative to the population analysis: visual predictive check (A), observations vs predictions (B), distribution of the individual weighted residuals with respect to time (C) and example of an individual fit (D) of the exponential (left), the logistic (center) and the Gompertz (right) models.

the models are ranked according to their AIC (Akaike information criterion). As shown below, the Gompertz model provided the lowest AIC values. Different types of model diagnostic plots are reported in Fig. 1. The visual predictive checks (VPCs) in Fig. 1A compare the empirical percentiles with the theoretical percentiles, *i.e.* those obtained from simulations of the calibrated models. Only in the case of the Gompertz model the observed percentiles were close to the predicted ones and remained within the corresponding prediction interval. The VPCs of the exponential and the logistic models exhibited model misspecification. The observations vs individual predictions of the Gompertz model in

Fig. 1B show a low percentage of outliers, *i.e.* the predictions outside of the 90% prediction interval. Moreover, the distribution of the observations were symmetrical around the predicted values with the Gompertz model (Fig. 1C) while the exponential and the logistic models provided skewed distributions. Figure 1D shows an example of individual fit with the three different models. This confirms that the Gompertz model describes better the dynamic of tumor growth.

Table 2 provides the values of the population parameters. The relative standard errors associated to population parameters were all low ($<10.7\%$), indicating good practical identifiability of the model parameters. Relative standard errors of the standard deviations of the random effects ω were all smaller than 34.1%.

Table 1. Models ranked in ascending order of AIC (Akaike information criterion). Other statistical indices are the log-likelihood estimate (-2LL) and the Bayesian information criterion (BIC). The reported values in the first row are the values of the indices of the best model (the Gompertz model). The other rows provide the difference of each statistical index between the model in the row and the Gompertz model. * The reduced Gompertz model is introduced in Sect. 3.2.

Model	-2LL	AIC	BIC
Gompertz	2232	2246	2253
Reduced Gompertz*	+24	+20	+18
Logistic	+83	+81	+80
Exponential	+412	+406	+403

3.2 The Reduced Gompertz Model

Correlation Between the Gompertz Parameters. Although the Gompertz parameters α and β were assumed to be independent, a high correlation within the population has been observed. Indeed, the SAEM algorithm estimated a correlation of the random effects equal to 0.957. Moreover, Fig. 2A shows the relation between the individual parameters, where we found $R^2 = 0.929$. This correlation was observed in other experimental system: in a rat mammary carcinoma system [16], human lung metastases from testicular tumors [9] and human benign tumors [17]. We write the relationship between the two parameters as:

$$\alpha^i = k\beta^i + c \tag{7}$$

where c is the intercept of the regression line, which is found close to zero. The slope of the regression line could be a characteristic constant of tumor growth within a certain species. From a biological point of view, this characteristic constant could be associated to the carrying capacity K, following the relation $K = V_0 \exp(k)$, where V_0 is the initial volume of the tumor. As previously remarked by [6], this result might be supported by the fact that a particular species is able to support a tumor of a certain maximum size.

Table 2. Fixed effects (typical values) of the parameters of the different models. CV = Coefficient of Variation, expressed in percentage and estimated as the standard deviation of the parameter divided by the fixed effect and multiplied by 100. σ is vector of the residual error model parameters. Last column shows the relative standard errors (R.S.E.) of the estimates. * The reduced Gompertz model is introduced in Sect. 3.2.

Model	Parameter	Unit	Fixed effects	CV (%)	R.S.E. (%)
Gompertz	α	day^{-1}	0.713	22.57	3.79
	β	day^{-1}	0.0731	318	5.77
	σ	–	[28.2, 0.081]	–	[13.8, 14.3]
Reduced Gompertz*	β	day^{-1}	0.0757	158.37	10.7
	k	–	9.51	–	5.26
	σ	–	[27.6, 0.106]	–	[14.03, 11.7]
Logistic	α	day^{-1}	0.477	25.48	2.84
	K	mm^3	1.65e+03	0.006	4.67
	σ	–	[38.5, 0.11]	–	[13.2, 14.01]
Exponential	α	day^{-1}	0.403	28.01	2.75
	σ	–	[87.8, 0.37]	–	[19.1, 14.8]

Biological Interpretation in Terms of the Proliferation Rate. By definition, the parameter α is equal to the specific growth rate at the time of injection. Assuming that the cells do not change their proliferation kinetics when implanted, this value should thus be equal to the *in vitro* proliferation rate (supposed to be the same for all the cells of the same cell line), denoted here by λ. The value of this biological parameter was assessed *in vitro* and found equal to 0.929 [3]. Confirming our theory, we indeed found estimated values of α close to λ (fixed effects of 0.713), although strictly smaller in the majority of the cases (Fig. 2A). This difference could be explained by the fact that not all the cells "take" when grafted in an animal. Denoting by $\hat{V}_0^i < V_0$ the volume of these cells, our assumption would rather be expressed as:

$$\lambda = \alpha^i - \beta^i \log \left(\frac{\hat{V}_0^i}{V_0} \right) > \alpha^i,$$

which was confirmed in our observations.

Population Analysis of the Reduced Gompertz Model. The high correlation among the Gompertz parameters, combined to the biological rationale explained above, suggested that a reduction of the degrees of freedom could improve identifiability of the parameters and yield a simpler model. Considering the relation in (7), and assuming c negligible, we thus propose the following reduced Gompertz model $V_R(t; \beta, k)$:

$$\frac{dV_R}{dt} = \beta^i k - \beta^i \log \left(\frac{V_R}{V_0} \right), \qquad i = 1, \dots, N \tag{8}$$

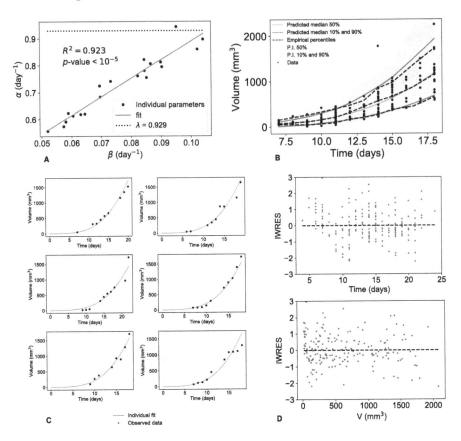

Fig. 2. Correlation between the parameters of the Gompertz model (A) and results of the population analysis of the reduced Gompertz model: visual predictive check (B), examples of individual fits (C) and scatter plots of the residuals (D).

where β has mixed effects, while k has only fixed effects, *i.e.* k is constant within the population.

Figure 2 shows the results relative to the population analysis performed with `Monolix`. We noticed a good description of the population (Fig. 2B) and of the individual trends (Fig. 2C) even if only one parameter has mixed effects. Moreover, the residuals are symmetrically distributed around zero (Fig. 2D).

Table 1 shows the statistical indices of the 1-d Gompertz model. Comparing these values with the other equations we noticed that the reduced model performes well compared to the other growth curves. Moreover, we obtained an excellent identifiability of the parameters (Table 2).

3.3 Prediction of the Time Since Tumor Initiation

We then studied the relative performances of the reduced Gompertz and the Gompertz models for the problem of predicting the initiation time from the three last measurements using Bayesian inference. For a given animal i, we consider as first observation $y^i_{n^i-2}$ and tried to predict $t^i_{n^i-2}$. Initial conditions were not considered equal to the number of injected cells anymore but rather to $y^i_{n^i-2}$. The value t^i_{pred} was defined as the time when the median value of the posterior predictive distribution of $\tilde{y}(u)$ reached V_0.

Different data sets were used for learning the priors (*training sets*) and for making predictions (*test sets*) by means of k-fold cross validation, with k equal to the total number of animals of the dataset ($k = N$). At each iteration we computed the parameters distribution of the population composed by $N - 1$ individuals and used this as prior to predict the initiation time of the excluded subject i. The **Stan** software was used to draw 2000 realizations from the posterior predictive distribution of the animal i. We eventually estimated the model accuracy (i.e. relative error of the prediction, defined by $\text{err}^i = t^i_{\text{pred}}/t^i_{n^i-2}$) and the uncertainty of the prediction (i.e. precision, measured by the width of the 90% prediction interval (PI)).

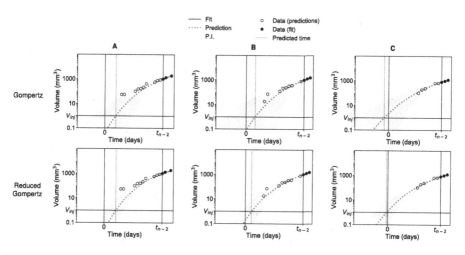

Fig. 3. Three examples of backward predictions of individuals A, B and C computed with Bayesian inference: Gompertz model (first row) and reduced Gompertz (second row). Only the last three points are considered to estimate the parameters. The grey area is the 90% prediction interval (P.I) and the dotted blue line is the median of the posterior predictive distribution. The red line is the predicted initiation time and the black vertical line the actual initiation time. (Color figure online)

Figure 3 shows some examples of prediction of three individuals and Fig. 4 shows the distribution of the relative error. The reduced Gompertz model was found to have better accuracy in predicting the initiation time (mean error = 9.4%) and to have the smallest uncertainty (mean precision = 7.34 days), while the Gompertz model had worse performances (mean error = 19.6% and precision = 18.2 days). Indeed the reduced Gompertz had only one parameter to estimate and the prior distribution allowed to have a reliable prediction.

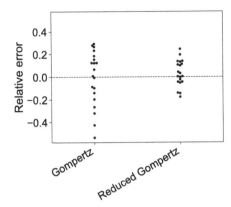

Fig. 4. Accuracy of the prediction models: swarmplots of relative errors of the Gompertz and the reduced Gompertz model.

4   Conclusions

We have performed a quantitative analysis of tumor growth kinetics using nonlinear mixed-effects modeling. This allowed us to propose a novel, "reduced" Gompertz model with one parameter less. We further developed a method for prediction of individual tumor age given few measurements. The approach is based on: (i) the application of the population approach in order to learn the parameter distribution of the models, (ii) the reduced Gompertz model with only one individual parameter and (iii) Bayesian inference to determine the posterior predictive distribution used to compute the time since initiation.

Our results warrant against the use of the exponential or logistic models for description of tumor growth, that were therefore excluded in the prediction of the age of a tumor. On the other hand, combining the population approach with a reduced version of the Gompertz model comprising one parameter only allows to reach a level of accuracy which offers promising clinical perspectives.

The method that we propose here remains to be extended to clinical data, although it will not be possible to have a firm confirmation since the entire natural history of neoplasms cannot be observed. Nevertheless, the encouraging results obtained here could allow to give approximate estimates. Such predictions could be informative in clinical practice to determine the extent of invisible metastatics at the time of diagnosis, by refining published methods [5].

References

1. Monolix version 2018R2. Lixoft SAS (2018)
2. Benzekry, S., et al.: Classical mathematical models for description and prediction of experimental tumor growth. PLoS Comput. Biol. **10**(8), e1003800 (2014). https://doi.org/10.1371/journal.pcbi.1003800

3. Benzekry, S., Tracz, A., Mastri, M., Corbelli, R., Barbolosi, D., Ebos, J.M.L.: Modeling spontaneous metastasis following surgery: an in vivo-in silico approach. Cancer Res. **76**(3), 535–547 (2016)

4. Bertram, J.S., Janik, P.: Establishment of a cloned line of Lewis Lung Carcinoma cells adapted to cell culture. Cancer Lett. **11**(1), 63–73 (1980)

5. Bilous, M., et al.: Quantitative mathematical modeling of clinical brain metastasis dynamics in non-small cell lung cancer. Sci. Rep. **9**(1) (2019). https://doi.org/10.1038/s41598-019-49407-3

6. Brunton, G.F., Wheldon, T.E.: Characteristic species dependent growth patterns of mammalian neoplasms. Cell Tissue Kinet **11**(2), 161–175 (1978)

7. Carpenter, B., et al.: Stan: a probabilistic programming language. J. Stat. Softw. **76**(1) (2017). https://doi.org/10.18637/jss.v076.i01

8. Collins, V.P., Loeffler, R.K., Tivey, H.: Observations on growth rates of human tumors. Am. J. Roentgenol. Radium Ther. Nucl. Med. **76**(5), 988–1000 (1956)

9. Demicheli, R.: Growth of testicular neoplasm lung metastases: tumor-specific relation between two Gompertzian parameters. Eur. J. Cancer **16**(12), 1603–1608 (1980). https://doi.org/10.1016/0014-2964(80)90034-1

10. Frenzen, C.L., Murray, J.D.: A cell kinetics justification for Gompertz' equation. SIAM J. Appl. Math. **46**(4), 614–629 (1986)

11. Gelman, A.: Bayesian Data Analysis. Chapman & Hall/CRC Texts in Statistical Science, 3rd edn. CRC Press, Boca Raton (2014)

12. Kramer, A., Calderhead, B., Radde, N.: Hamiltonian Monte Carlo methods for efficient parameter estimation in steady state dynamical systems. BMC Bioinform. **15**(1), 253 (2014). https://doi.org/10.1186/1471-2105-15-253

13. Laird, A.K.: Dynamics of tumor growth. Br. J. Cancer **13**, 490–502 (1964)

14. Lavielle, M.: Mixed Effects Models for the Population Approach: Models, Tasks, Methods and Tools. Chapman & Hall/CRC Biostatistics Series. Taylor & Francis, Boca Raton (2014)

15. Norton, L.: A Gompertzian model of human breast cancer growth. Cancer Res. **48**(24), 7067–7071 (1988)

16. Norton, L., Simon, R., Brereton, H.D., Bogden, A.E.: Predicting the course of Gompertzian growth. Nature **264**(5586), 542–545 (1976). https://doi.org/10.1038/264542a0

17. Parfitt, A.M., Fyhrie, D.P.: Gompertzian growth curves in parathyroid tumours: further evidence for the set-point hypothesis. Cell Prolif. **30**(8–9), 341–349 (1997)

18. Steel, G.G.: Growth Kinetics of Tumours: Cell Population Kinetics in Relation to the Growth and Treatment of Cancer. Clarendon Press, Oxford (1977)

19. Steel, G.G.: Species-dependent growth patterns for mammalian neoplasms. Cell Tissue Kinet **13**(4), 451–453 (1980)

20. Vaghi, C., et al.: A reduced Gompertz model for predicting tumor age using a population approach. bioRxiv (2019). https://doi.org/10.1101/670869

21. Winsor, C.P.: The Gompertz curve as a growth curve. Proc. Natl. Acad. Sci. U.S.A. **18**(1), 1–8 (1932)

Author Index

Printed in the United States
By Bookmasters